AF849070

SYSTEMS ENGINEERING METHODS, DEVELOPMENTS AND TECHNOLOGY

FAULT DETECTION

METHODS, APPLICATIONS AND TECHNOLOGY

Systems Engineering Methods, Developments and Technology

Additional books in this series can be found on Nova's website under the Series tab.

Additional e-books in this series can be found on Nova's website under the e-book tab.

SYSTEMS ENGINEERING METHODS, DEVELOPMENTS AND TECHNOLOGY

FAULT DETECTION

METHODS, APPLICATIONS AND TECHNOLOGY

DANIEL MARTIN
EDITOR

nova publishers
New York

Copyright © 2017 by Nova Science Publishers, Inc.

All rights reserved. No part of this book may be reproduced, stored in a retrieval system or transmitted in any form or by any means: electronic, electrostatic, magnetic, tape, mechanical photocopying, recording or otherwise without the written permission of the Publisher.

We have partnered with Copyright Clearance Center to make it easy for you to obtain permissions to reuse content from this publication. Simply navigate to this publication's page on Nova's website and locate the "Get Permission" button below the title description. This button is linked directly to the title's permission page on copyright.com. Alternatively, you can visit copyright.com and search by title, ISBN, or ISSN.

For further questions about using the service on copyright.com, please contact:
Copyright Clearance Center
Phone: +1-(978) 750-8400 Fax: +1-(978) 750-4470 E-mail: info@copyright.com.

NOTICE TO THE READER

The Publisher has taken reasonable care in the preparation of this book, but makes no expressed or implied warranty of any kind and assumes no responsibility for any errors or omissions. No liability is assumed for incidental or consequential damages in connection with or arising out of information contained in this book. The Publisher shall not be liable for any special, consequential, or exemplary damages resulting, in whole or in part, from the readers' use of, or reliance upon, this material. Any parts of this book based on government reports are so indicated and copyright is claimed for those parts to the extent applicable to compilations of such works.

Independent verification should be sought for any data, advice or recommendations contained in this book. In addition, no responsibility is assumed by the publisher for any injury and/or damage to persons or property arising from any methods, products, instructions, ideas or otherwise contained in this publication.

This publication is designed to provide accurate and authoritative information with regard to the subject matter covered herein. It is sold with the clear understanding that the Publisher is not engaged in rendering legal or any other professional services. If legal or any other expert assistance is required, the services of a competent person should be sought. FROM A DECLARATION OF PARTICIPANTS JOINTLY ADOPTED BY A COMMITTEE OF THE AMERICAN BAR ASSOCIATION AND A COMMITTEE OF PUBLISHERS.

Additional color graphics may be available in the e-book version of this book.

Library of Congress Cataloging-in-Publication Data

ISBN 978-1-53610-345-8

Published by Nova Science Publishers, Inc. † New York

CONTENTS

PREFACE

Fault detection and identification (FDI) play a fundamental role in most modern industrial systems and processes. They are essential in ensuring safe and reliable operation. In this book, Chapter One and Chapter Two discuss general systems and the use of linearizations for active fault detection in nonlinear differential algebraic equations. Chapter Three provides a comparative analysis and comprehensive review of the existing fault detection methods, including manual picking and computer-aided automatic/semi-automatic extraction. Chapter Four studies self-adaptive expert systems for process monitoring and fault detection. The final chapter provides a review of an integrated Fault Detection and Diagnostics (FDD) system for HVAC&R based on virtual sensors.

As explained in Chapter 1, fault detection and identification are important tasks in most modern industrial systems and processes. A variety of approaches, both active and passive have been investigated in the literature. Many complex systems are most naturally modeled by differential algebraic equations (DAEs). Most of this prior work assumed the models were not DAEs and if the models were DAEs, then the DAEs were linear index one. Recently, new results have appeared on auxiliary signal design for fault detection in linear differential algebraic equations. Chapter 1 examines using this approach directly on a nonlinear DAE. Chapter 1 begins with a general nonlinear discussion and then carries out a nontrivial case study on a nonlinear index three robotic arm model.

Fault detection and identification (FDI) play a fundamental role in most modern industrial systems and processes. They are essential in ensuring safe and reliable operation. A variety of approaches, both active and passive have been investigated. Most of that work required the models to be linear and modeled by ordinary differential equations. Recently, new results on auxiliary signal design for fault detection in linear differential algebraic equations (DAEs) has led to a technique that can potentially be used on some linear DAE models. Chapter 2 examines this application with a nontrivial case study involving a nonlinear robot arm model. It is shown the previous theory does not immediately apply to the linearizations and some modifications are sometimes necessary in order to use this approach on the linearized models that occur in some applications. Techniques for reformulating are presented. It is also shown how to evaluate a test signal for a nonlinear problem found by this or any other method.

As shown in Chapter 3, in petroleum exploration and production, faults are of importance by controlling pathways of hydrocarbon migration and accumulation in the subsurface, and robust fault detection is one of the major tasks of geologic and geophysical interpretation from three-dimensional (3D) seismic data. Traditionally, faults are interpreted by manually picking on vertical/horizontal seismic sections with geologic consistence; however, such manual tool is time consuming and sensitive to the visibility and interpreters' bias, especially for a large dataset with structural complexities. With the development of new signal processing and data visualization technologies, computer-aided semi-automatic/automatic fault extraction has been the focus of recent geophysical research on seismic fault detection with superiorities in both computational efficiency and result accuracy. Various methods have been developed and implemented in the petroleum industry; however, their results are often limited by the detection resolution and contaminated by non-fault artifacts that interfere with reliable structural interpretation. Chapter 3 provides a comparative analysis and comprehensive review of the existing fault detection methods, including manual picking and computer-aided automatic/semi-automatic extraction. A suite of algorithms for each method are tested regarding their computational efficiency and result accuracy by calibrating their results to the original seismic images for quality control. Finally, for addressing the limitations and problems observed in the existing techniques, innovative solutions are presented to further enhance the reliability of fault interpretation from seismic data.

Process monitoring and early fault detection are paramount in modern industries. As explained in Chapter 4 a well-developed process monitoring system is expected to reduce failure rates, production costs, and prevent health and environmental hazards. An abundance of sophisticated monitoring and fault detection strategies is available, most of which require either accurate process models, including models of the most frequently occurring faults, or large databases filled with relevant historical data. Unfortunately, neither of the previous is usually available, and process monitoring is often reduced to more-or-less elaborate limit-checking. Even in the case when the models and/or historical datasets are available, the obtained monitoring techniques are highly specialized and must usually be re-developed for any new application at hand. Chapter 4 presents a general-purpose, self-adaptive expert system for online and offline process monitoring with possibility of fault detection, capable to learn the properties of the process being monitored and to identify and separate its working regimes, both nominal and unwanted ones. The proposed system is also capable to identify changes in the process operation; both abrupt and incipient ones. The flexibility of the system is achieved by separating the monitoring algorithm into two distinct layers. The first layer, which is application dependent, is called Feature Generator (FG). FG uses current measurements of process variables to compute "features," numerical values used to concisely describe current working conditions. Features are then fed to the second layer, Potential Based Recursive Clustering (PBRC) algorithm, an original, novel adaptive clustering procedure. In essence, PBRC is designed as self-learning and auto-growing neural-network which tracks, in a robust manner, the motion of the process in the features space. It identifies, in an adaptive fashion, areas corresponding to separate process working conditions and decides which of the working conditions is currently active. If a novel working regime is encountered, PBRC can enlarge its knowledge base and incorporate the novelty for further reference. The system was tested and validated on a number of laboratory installations and also in real industrial environment. Obtained results showed that proposed system can detect faults and changes in working regime in the early stages with high accuracy and reliability.

According to the US Department of Energy, space heating, ventilation and air conditioning system account for 40% of residential primary energy use and for 30% of primary energy use in commercial buildings. A study released by the Energy Information Administration indicated that packaged air conditioners are widely used in 46% of all commercial buildings in the US Chapter 5 indicates that the annual cooling energy consumption related to the packaged air conditioner is about 160 trillion Btus. Therefore, an automated FDD (Fault Detection and Diagnostics) system that can automatically detect and diagnose faults and evaluate fault impacts has the potential for improving energy efficiency along with reducing service costs and comfort complaints. The primary bottlenecks to diagnostic implementation in the field are the high initial costs of additional sensors. To prevent those limitations, virtual sensors with low cost measurements and simple models are developed to estimate quantities that would be expensive and or difficult to measure directly. The use of virtual sensors can reduce costs compared to the use of real sensors and provide additional information for economic assessment. The virtual sensor can be embedded in a permanently installed control or monitoring system and continuous monitoring potentially leads to early detection of faults. The virtual sensors of individual equipment components can be integrated to estimate overall diagnostic information using the output of each virtual sensor.

In: Fault Detection
Editor: Daniel Martin
ISBN: 978-1-53610-345-8
© 2017 Nova Science Publishers, Inc.

Chapter 1

Active Fault Detection in Nonlinear Differential Algebraic Equations I: General Systems

Stephen L. Campbell*
Department of Mathematics,
North Carolina State University,
Raleigh, North Carolina, US

Abstract

Fault detection and identification are important tasks in most modern industrial systems and processes. A variety of approaches, both active and passive have been investigated in the literature. Many complex systems are most naturally modeled by differential algebraic equations (DAEs). Most of this prior work assumed the models were not DAEs and if the models were DAEs, then the DAEs were linear index one. Recently, new results have appeared on auxiliary signal design for fault detection in linear differential algebraic equations. This paper examines using this approach directly on a nonlinear DAE. The paper begins with a general nonlinear discussion and then carries out a nontrivial case study on a nonlinear index three robotic arm model.

Keywords: Fault detection, signal design, nonlinear differential algebraic equations

AMS Subject Classification: 53D, 37C, 65P

*E-mail address: slc@ncsu.edu

1. Introduction

In a multi-model active approach for fault detection we assume there are known models for the nominal and faulty behaviors of the system. The aim of the auxiliary signal is to facilitate detection by forcing the input-output sets for the nominal and faulty models to be disjoint. An additional goal in auxiliary signal design is to find the smallest such signal that guarantees detection, thereby limiting the effect of the signal on the system. Active detection has been studied extensively for systems modeled by ordinary differential equations (ODEs) in [8, 25, 26, 27, 28]. The closely related problem about input selection in observer design is studied in [4, 31, 30]. Some nonlinear results are in [1, 15]. Linear differential algebraic equation (DAE) models are considered in [34]. Using observers for fault detection in linear DAE models is discussed in [32, 33].

Fault detection or identification for DAEs has been considered by a number of authors. We note only [2, 11, 12, 16, 18, 19, 20, 24, 37, 39]. However, that work was generally limited to linear and index one DAEs. In contrast, the approach of [34] can be used for some higher index linear DAE models. This paper shows for the first time that approach can be extended to an important class of higher index nonlinear DAE models. It is assumed that the reader has a basic understanding of DAEs and their index [6, 23].

Section 2 will quickly review the ideas of [34, 35] in order to show what is different about this paper. Our nonlinear approach is introduced in Section 3 Applications are not always initially formulated so that they immediately meet the assumptions of Section 3 Section 4 discusses how some problems may be better formulated. Section 5 carefully discusses several issues in the context of a nonlinear control problem. The conclusion is in Section 6.

This paper focuses on the finding of a proper test signal. The companion paper [9] discusses the evaluation and use of a test signal on a nonlinear DAE. The signals found in [9] are found very differently from the techniques of this paper by using linearizations. However, the evaluation and application of a test signal are the same whether the test signal is found with the techniques in this paper or the techniques in [9] or by some other method so we will not repeat that discussion here.

It is important to note that while [9] and this paper start with the same robot arm model, the ensuing analysis and specific questions asked are quite different. In this paper we assume that system is at a set point consistent with the fault and a test signal is essential to find the fault. We also introduce the noise terms at the start rather than after a linearization. Furthermore, a different modified problem is derived.

2. Review of Linear DAE Approach

In this section we quickly review some ideas from [34] to frame our nonlinear discussion. Consider a multi-model formulation with two models and additive uncertainty. One model is the nominal system ($i = 0$) and the other is the faulty system ($i = 1$). Each model has the form,

$$E_i x_i' + F_i x_i = B_i u + M_i \mu_i \tag{1a}$$
$$y_i = C_i x_i + D_i u + N_i \eta_i. \tag{1b}$$

Here u is the test signal to be applied. Matrices E_i, F_i, B_i, M_i, C_i, D_i, N_i are constant in the original models. E_i may be singular allowing, in general, the models to be DAEs. Outputs y_0, y_1 must have the same dimension, but x_0, x_1 do not have to have the same dimension, allowing structural faults. The ν_i, η_i are additive uncertainties representing noise or model error. There are several variations of this problem such as including known exogenous input and feedthrough terms that can be easily handled given the information provided in this paper. The N_i are assumed nonsingular. That is, we allow noise or uncertainty in all equations.

For each model the uncertainty consists of $x_i(0)$, μ_i, ν_i and is measured by

$$\Gamma_i^2(x_i(0), \mu_i, \nu_i) = \frac{1}{2} x_i(0)^T P_i x_i(0) + \frac{1}{2}\int_0^T \|\nu_i\|^2 + \|\mu_i\|^2 \, dt. \tag{2}$$

Two ways of measuring the total uncertainty in both models have been considered. We will use

$$\Gamma_2^2 = \Gamma_0^2 + \Gamma_1^2 \tag{3}$$

which was introduced in [10]. We will assume that the total amount of uncertainty is bounded by γ so that $\Gamma_2^2 \leq \gamma^2$.

We assume that the test is to be performed over an interval $0 \leq t \leq T$. A test signal u is proper if it is not possible to get the same output from both models and still have the total uncertainty bound hold. That is, the smallest amount of uncertainty for which $y_1 = y_0$ violates the total uncertainty bound.

We seek the smallest proper test signal. There are several ways to measure the size of the test signal [8]. We shall use the energy or L^2 bound,

$$\|u\|^2 = \int_0^T \|u(t)\|^2 dt. \tag{4}$$

Use of a different measure of u is usually trivial to implement since it often amounts to altering one line of code where the outer cost is given. Several examples are given in [8].

Since we want the minimal proper signal, we must solve

Problem 1.

$$\textit{Find } J = \inf_u ||u||^2 \tag{5a}$$

$$\textit{such that}$$

$$\left(\inf_{\substack{x_i(0),\mu_i,\eta_i \\ i=0,1}} \Gamma^2(u) \right) \geq \gamma^2 \tag{5b}$$

$$E_0 x_0' + F_0 x_0 = B_0 u + M_0 \mu_0 \tag{5c}$$

$$E_1 x_1' + F_1 x_1 = B_1 u + M_1 \mu_1 \tag{5d}$$

$$0 = y_0 - y_1. \tag{5e}$$

The solution of Problem 1 is known as the minimal proper u with respect to the L^2 bound. Frequently, for linear models, the minimal proper u is unique up to a factor of negative one. However, examples can be constructed where it is not unique. From [34], a reformulation of the problem, making it easier to solve numerically is gotten by replacing the inner minimization (5b) with the necessary conditions for this minimization and also adding a collocation to compute the cost of the inner minimization.

Fully describing how Problem 1 can be reformulated and explicitly giving all the transformations would run for several pages. Full details can be found in [34]. The necessary conditions for the inner minimum which is the left hand side of (5b) are taken as constraints when solving the original problem (5a). The development in this paper is similar in spirit but differs in a number of details. Our goal here is to work more directly with Problem 1 in the nonlinear case.

It is important to get a formulation that can be implemented in high quality direct transcription optimal control codes such as SOCX [5] or

GPOPS-II [29]. SOCX is available from the Applied Mathematical Analysis, `www.appliedmathematicalanalysis.com/`. In this paper, the simulations are generated from the Matlab based GPOPS-II. Both SOCX and GPOPS-II have two options when solving the nonlinear programming problem (NLP) that results from the discretizations. SOCX comes with the interior point method IPOPT [38] and the sequential quadratic programming method SNOPT [21]. GPOPS-II only comes with IPOPT but a separate license for SNOPT is available for a fee. The results in this paper relied on IPOPT.

3. General Nonlinear Approach

In extending the results of the linear case to nonlinear DAE models several changes are necessary. One is that the developed formulations need to be solvable by state of the art numerical software. Secondly, some of the operations used in the linear case would now involve nonlinear coordinate changes and thus while theoretically available may need to be avoided computationally and algorithmically.

We shall first discuss in general what is needed for the approach to work on a nonlinear DAE problem. Then we will consider an important specific class of problems where we can provide more detail. Finally we shall discuss a specific complex nonlinear problem. To simplify the discussion we shall assume additive uncertainty. While there is some software that can work with general fully implicit DAEs, this software is not generally available. We limit ourselves to a wide class of nonlinear systems which are amenable to modern software. We assume we have the two models and output equations of the form

$$x_i' = F_i(x_i, a_i, u, t) + M_i\mu_i \tag{6a}$$

$$0 = g_i(x_i, a_i, t, u) + R_i\eta_i \tag{6b}$$

$$y_i = G_i(x_i, a_i, t, u) + N_i\nu_i \tag{6c}$$

Here the DAE model is (6a) and (6b) and the output equation is (6c). The x_i variable from (1) has been divided into two variables x_i, a_i. We assume that the outputs y_0, y_1 have the same dimension. The x_0, a_0 can have different dimensions from the x_1, a_1. Here the a_i represent algebraic variables which do not appear differentiated. The x_i are called diferential variables. Note that we do not assume that either $\partial g_i/\partial a_i$ is nonsingular, so we are not assuming that the models are index one. Thus there can be additional hidden constraints besides the g_i. We assume that N_i is nonsingular. This means that we allow noise in all output channels and is not a restrictive assumption for finding a proper test signal. Depending on the application, R_i may be nonsingular or it may be zero. The second case arises with physical constraints and will be utilized in the next section. We also assume that the noises are independent which again only increases the robustness of our test signal. Since the inner optimization will set equal to zero any parts of the noise in the null space of M_i we may assume that M_i is full column rank.

For the nonlinear control problem, if the control set point $\bar{u}$ is nonzero we can just use $u - \bar{u}$ in the test signal cost since $u - \bar{u}$ is the test signal. For the state set point $\bar{x}$, the simplest way way to incorporate this is to view the noise in the initial condition as $\bar{x} - x(0) = \delta x(0)$ and we will do so on the nonlinear problem.

The general inner minimization problem is to minimize

$$\sum_{i=0}^{1} \frac{1}{2}\left(\delta x_i(0)^T P_i \delta x_i(0) + \int_0^T \|\mu_i\|^2 + \|\nu_i\|^2 + \|\eta_i\|^2 dt\right) \tag{7a}$$

subject to

$$\begin{aligned}
x_0' &= F_0(x_0, a_0, t, u) + M_0\mu_0 && (7b)\\
x_1' &= F_1(x_1, a_1, t, u) + M_1\mu_1 && (7c)\\
0 &= g_0(x_0, a_0, u, t) + R_0\eta_0 && (7d)\\
0 &= g_1(x_1, a_1, u, t) + R_1\eta_1 && (7e)\\
0 = y_0 - y_1 &= G_0(x_0, a_0, u, t) + N_0\nu_0 - G_1(x_1, a_1, u, t) - N_1\nu_1 && (7f)
\end{aligned}$$

There is a special case, discussed later in the robotic example, where the system is operating at a set point x_s, u_s and it is a set point for both the nominal and the faulty systems. The test signal is essential in order to determine if there is a fault. In this case $P_i = \infty I$, and the boundary conditions at the left end point are just $x_i(0) = x_s$.

Note that while we can take the N_i invertible, that the R_i might be zero if they represent physical constraints. The derivative of a vector valued function k of vector variable w is denoted k_w. If

$$J = \begin{bmatrix} g_{0,a_0} & 0 \\ 0 & g_{1,a_1} \end{bmatrix} \tag{8}$$

is nonsingular, then (7b)–(7f) is an index one DAE in the x_i, a_i for a given u, and some equations can be eliminated from the necessary conditions as illustrated below. In general, given vectors z_0, z_1 we let z have column entries z_0, z_1. Thus, for example,

$$\nu = \begin{bmatrix} \nu_0 \\ \nu_1 \end{bmatrix}, \eta = \begin{bmatrix} \eta_0 \\ \eta_1 \end{bmatrix}, \mu = \begin{bmatrix} \mu_0 \\ \mu_1 \end{bmatrix}.$$

However, if J is singular, then one can proceed as follows. Suppose that

$$\tilde{J} = \begin{bmatrix} g_{0,a_0} & 0 \\ 0 & g_{1,a_1} \\ G_{0,a_0} & G_{1,a_1} \end{bmatrix} \tag{9}$$

is full column rank and

$$\hat{J} = \begin{bmatrix} g_{0,a_0} & 0 & R_0 & 0 & 0 & 0 \\ 0 & g_{1,a_1} & 0 & R_1 & 0 & 0 \\ G_{0,a_0} & G_{1,a_1} & 0 & 0 & N_0 & N_1 \end{bmatrix} \tag{10}$$

is full row rank.

There there is a subset of the variables $\{\nu, \eta\}$, so that the Jacobian of the constraints (7d)–(7f) with respect to these variables is nonsingular. We do not need to actually find this set of variables, it is enough to know it exists [7, 17]. In this case the Lagrangian of the inner minimization is

$$\begin{aligned} \mathcal{L} \;=\; & \|\mu\|^2 + \|\eta\|^2 + \|\nu\|^2 \\ & +\lambda_0^T \left(F_0(x_0, a_0, t, u) + M_0\mu_0\right) + \lambda_1^T \left(F_1(x_1, a_1, t, u) + M_1\mu_1\right) \\ & +\lambda_2 \left(g_0(x_0, a_0, u_0, t) + R_0\eta_0\right) + \lambda_3^T \left(g_1(x_1, a_1, u, t) + R_1\eta_1\right) \\ & \lambda_4^T \left(G_0(x_0, a_0, u, t) + N_0\nu_0 - G_1(x_1, a_1, u, t) - N_1\nu_1\right) \end{aligned} \tag{11}$$

Then the necessary conditions for the inner minimization are

$$x_0' = F_0(x_0, a_0, t, u) + M_0\mu_0 \tag{12a}$$
$$x_1' = F_1(x_1, a_1, t, u) + M_1\mu_1 \tag{12b}$$
$$0 = g_0(x_0, a_0, u, t) + R_0\eta_0 \tag{12c}$$
$$0 = g_1(x_1, a_1, u, t) + R_1\eta_1 \tag{12d}$$

$$
\begin{aligned}
0 &= G_0(x_0, a_0, u, t) + N_0\nu_0 - G_1(x_1, a_1, u, t) - N_1\nu_1 && \text{(12e)}\\
\lambda_0' &= F_{0,x_0}(x_0, a_0, t, u)^T\lambda_0 + +g_{0,x_0}^T\lambda_2 + 2G_{0,x_0}(x_0, a_0, u, t)\lambda_4 && \text{(12f)}\\
\lambda_1' &= F_{1,x_1}(x_1, a_1, t, u)^T\lambda_1 + g_{1,x_1}^T\lambda_3 + 2G_{1,x_1}(x_1, a_1, u, t)\lambda_4 && \text{(12g)}\\
0 &= 2\mu_0 + M_0^T\lambda_0 && \text{(12h)}\\
0 &= 2\mu_1 + M_1^T\lambda_1 && \text{(12i)}\\
0 &= 2\nu_0 + N_0^T\lambda_4 && \text{(12j)}\\
0 &= 2\nu_1 - N_1^T\lambda_4 && \text{(12k)}\\
0 &= 2\eta_0 + R_0^T\lambda_2 && \text{(12l)}\\
0 &= 2\eta_1 + R_1^T\lambda_3 && \text{(12m)}
\end{aligned}
$$

Along with the boundary conditions

$$
\begin{aligned}
\lambda_0(0) &= -P_0\delta x_0(0) && \text{(12n)}\\
\lambda_1(0) &= -P_1\delta x_1(0) && \text{(12o)}\\
\lambda_0(T) &= 0 && \text{(12p)}\\
\lambda_1(T) &= 0 && \text{(12q)}
\end{aligned}
$$

In addition there is the dynamics and cost constraint on the inner minimization solution to guarantee that u is proper.

$$
\begin{aligned}
\psi' &= \|\mu\|^2 + \|\nu_0\|^2 + \|\eta\|^2 + \|\nu\|^2 && \text{(12r)}\\
\psi(0) &= \frac{1}{2}\left(x_0(0)^T P_0\delta x_0(0) + \delta x_1(0)^T P_1\delta x_1(0)\right) && \text{(12s)}\\
\psi(T) &\geq \gamma^2. && \text{(12t)}
\end{aligned}
$$

In the case where the initial point x_s is known, (12s) becomes $\psi(0) = 0$ and (12n) and (12o) become $x_0(0) = x_s$ and $x_1(0) = x_s$.

The resulting optimization problem is to

$$
\begin{gathered}
\min \|u\|^2 \\
\text{such that equations (12) hold} \qquad \text{(13)}
\end{gathered}
$$

It has been our experience with solving DAE optimization problems numerically that it is helpful to reduce the number of algebraic constraints if possible. For example, here (12h)–(12m) can be used to move some algebraic constraints from being constraints seen by the NLP solver to just being part of the evaluation of the cost.

4. Hessenberg Systems

DAEs from applications may not always be in a form suitable for immediately applying this test signal design approach and some reformulation may be needed. That reformulation is discussed in this section. To make the discussion general and yet specific enough to lend insight we shall now assume that models are index three DAEs in Hessenberg form [6]. Index one DAEs and index two DAEs in Hessenberg form are handled in a similar manner but are simpler to work with.

In this subsection we shall assume that the models are initially in the form

$$x_i' = f_i(x_i, z_i, w_i, u) + F_i\mu_i \tag{14a}$$
$$z_i' = g_i(x_i, z_i, u, t) + G_i\nu_i \tag{14b}$$
$$0 = h_i(z_i, t) + H_i\zeta_i \tag{14c}$$

with output

$$y_i = q_i(x_1, z_i, w_i, t) + Q_i\gamma_i. \tag{14d}$$

The x_i in (6) is now x_i, z_i and the a_i from (6) is now w_i. This change of notation in this section is to make the presentation similar to that of the usual discussions of Hessenberg systems in the literature. The uncertainties are $\mu_i, \nu_i, \zeta_i, \gamma_i$.

In working with the inner optimization problem we need to be able to put the model into a form, using doable transformations, from which we can derive necessary conditions and be able to solve these necessary conditions numerically.

The inner minimization problem is to minimize

$$\frac{1}{2}\delta x(0)^T P \delta x(0) + \frac{1}{2}\int_0^T \|\mu\|^2 + \|\nu\|^2 + \|\zeta\|^2 + \|\gamma\|^2 dt \tag{15}$$

Subject to

$$x_0' = f_0(x_0, z_0, w_0, u) + F_0\mu_0 \tag{16a}$$
$$x_1' = f_1(x_1, z_1, w_1, u) + F_1\mu_1 \tag{16b}$$
$$z_0' = g_0(x_0, z_0, u, t) + G_0\nu_0 \tag{16c}$$
$$z_1' = g_1(x_1, z_1, u, t) + G_1\nu_1 \tag{16d}$$
$$0 = h_0(z_0, t) + H_0\zeta_0 \tag{16e}$$
$$0 = h_1(z_1, t) + H_1\zeta_1 \tag{16f}$$
$$0 = q_0(x_0, z_0, w_0, t) + Q_0\gamma_0 - q_1(x_1, z_1, w_1, t) - Q_1\gamma_1 \tag{16g}$$

In solving the inner minimization problem, u is from the outer minimization. The x_i, z_i are state variables and the μ_i, ν_i, γ_i, ζ_i, w_i are algebraic variables. For most software, the inner minimization equations are easier to work with if they are index one. A careful examination of several formulations and how to do the reformulation is given in [22]. There are two separate things that can be done to reduce the index. One is to reduce the index of the individual models, the second is by modifying the cost so that a high index algebraic variable appears in the cost. This will reduce the index of the variational equations even if the individual models do not have a reduced index.

To reduce the index of an individual model the simplest way to do this is by differentiating a constraint and then dropping an equation. However, the differentiating of an unknown disturbance or a numerically computed quantity is not desirable. Fortunately in many applications there is disturbance or noise in only some of the equations. Some equations describe physical laws or structural relationships such as current or voltage laws and do not have noise terms. Accordingly we will assume that (16) has the form

$$x_0' = f_0(x_0, z_0, w_0, u) + F_0\mu_0 \tag{17a}$$
$$x_1' = f_i(x_1, z_1, w_1, u) + F_1\mu_1 \tag{17b}$$
$$z_0' = g_0(x_0, z_0, t) \tag{17c}$$
$$z_1' = g_1(x_1, z_1, t) \tag{17d}$$
$$0 = h_0(z_0, t) \tag{17e}$$
$$0 = h_1(z_1, t) \tag{17f}$$
$$0 = q_0(x_0, z_0, w_0, t) - Q_0\gamma_0 - q_1(x_1, z_1, w_1, t) - Q_1\gamma_1 \tag{17g}$$

We will see in the case study section that these assumptions are reasonable.

We may assume that the Q_i are nonsingular. Then differentiating (17e) and (17f) once analytically with respect to t and eliminating the differentiated variables and then dropping the derivatives (17c)-(17d) gives

$$x_0' = f_0(x_0, z_0, w_0, u) + F_0\mu_0 \tag{18a}$$
$$x_1' = f_i(x_1, z_1, w_1, u) + F_1\mu_1 \tag{18b}$$
$$0 = h_{0,z_0} g_0(x_0, z_0, t) + h_{0,t} \tag{18c}$$
$$0 = h_{1,z_1} g_1(x_1, z_1, t) + h_{1,t} \tag{18d}$$

$$0 = h_0(z_0, t) \tag{18e}$$
$$0 = h_1(z_1, t) \tag{18f}$$
$$0 = q_0(x_0, z_0, w_0, t) + Q_0\gamma_0 - q_1(x_1, z_1, w_1, t) - Q_1\gamma_1 \tag{18g}$$

Note that this operation has turned the differential variables z_0, z_1 into algebraic variables but they now appear in the algebraic constraints. Whether to drop all or part of (17c)–(17d) depends on the rank of h_{0,z_0} and h_{1,z_1}. We have done the case here where they are invertible. For the necessary conditions to be index one, we need the Jacobian of all the algebraic constraints and the cost with respect to all algebraic variables to be nonsingular. If necessary, additional reductions can sometimes be used. However, this becomes more complex. The exact steps to be carried out and what the desired model is will be dependent on the particular example and technique used. Note that in this paper and in [9] different modified models are derived.

5. Case Study

To further examine the ideas of the previous sections, we consider the model of a robot arm with elastic joints which is shown in Figure 1. Robot systems often use harmonic drives, belts, or long shafts as transmission elements between the motors and the links (arms) and typically display oscillations both in fast motion and after a sudden stop. Experimental tests and simulations have shown that the elasticity introduced at the joints by these transmission elements is the major reason for their vibrational behavior, so we must include elasticity in the dynamic model. As a result, the internal position of the motors does not determine the position of the driven arms. The dynamics of this displacement can be modeled by inserting a linear torsional spring at each elastic joint between the actuator and the link. The modeling process is described in full in [13].

The basic model given here is the same as that of [9] but here we define a different fault scenario. We also derive a different modified model and discuss some different issues in working with DAEs. Here we shall assume that the system is operating at a set point x_s. Without loss of generality we may assume the set point control is $u_s = 0$. It is assumed that x_s, u_s is a also a set point for the faulty model. Thus some controller action is needed to reveal the fault if it is present. We assume that it is desired to keep end point motion vertical during the test.

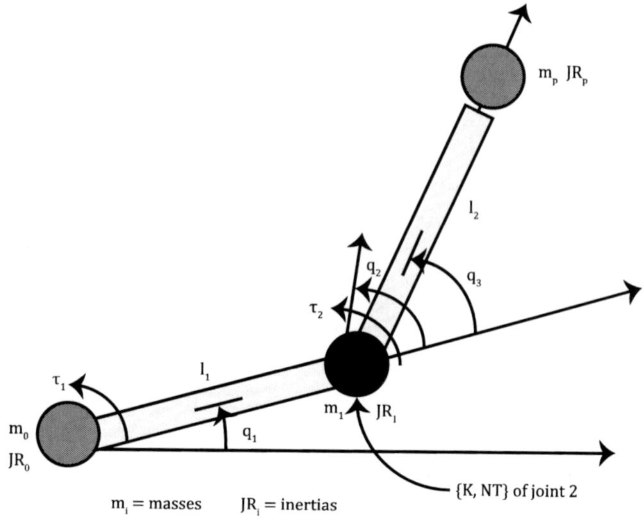

Figure 1. A two-link robot arm with the second joint elastic, from [14].

The model for the robot arm shown in Figure 1 is given by (19). m_0, m_1, and m_p are masses, with m_p denoting the load or object being held, l_1 and l_2 are the lengths of the arms, K is the coefficient of elasticity of joint 2, NT is the transmission ratio at the second joint, JR_i are rotor inertias, q_i are angular coordinates describing the robot's configuration, and τ_i are the rotational torques caused by the drive motors. In this model $x = [q_1, q_2, q_3, q_1', q_2', q_3']^T$, $u_1 = \tau_1$, and $u_2 = \tau_2$.

$$x_1' = x_4 \tag{19a}$$
$$x_2' = x_5 \tag{19b}$$
$$x_3' = x_6 \tag{19c}$$
$$x_4' = f_4(x_2, x_3, x_4, x_6) + g_{41}(x_3)u_1 - g_{41}(x_3)u_2 + \mu_1 \tag{19d}$$
$$x_5' = f_5(x_2, x_3, x_4, x_6) - g_{41}(x_3)u_1 + g_{52}(x_3)u_2 + \mu_2 \tag{19e}$$
$$x_6' = f_6(x_2, x_3, x_4, x_6) + g_{61}(x_3)u_1 - g_{61}(x_3)u_2 + \mu_3 \tag{19f}$$
$$0 = \cos x_1 + \cos(x_1 + x_3) - 1 \tag{19g}$$

Equation (19g) is a path constraint describing vertical motion on the endpoint of the robotic arm. The nonlinear functions are

$$g_{41}(x_3) = \frac{A_2}{A_3(A_4 - A_3\cos^2 x_3)}$$
$$g_{52}(x_3) = g_{41}(x_3) + \frac{1}{JR_1}$$
$$g_{61}(x_3) = -g_{41}(x_3) - \frac{\cos x_3}{A_4 - A_3\cos^2 x_3},$$

$$\begin{aligned}
f_4(x_2,x_3,x_4,x_6) &= \frac{A_2\sin x_3(x_4+x_6)^2 + A_3x_4^2\sin x_3\cos x_3}{A_4 - A_3\cos^2 x_3} \\
&\quad + \frac{K\left(x_3 - \frac{x_2}{NT}\right)\left(\frac{A_2}{A_3}\left(\frac{NT-1}{NT}\right) + \cos x_3\right)}{A_4 - A_3\cos^2 x_3}, \\
f_5(x_2,x_3,x_4,x_6) &= -f_4(x_2,x_3,x_4,x_6) + \frac{K}{NT}\left(x_3 - \frac{x_2}{NT}\right)\left(\frac{1}{JR_1} - 2g_{41}(x_3)\right), \\
f_6(x_2,x_3,x_4,x_6) &= -f_4(x_2,x_3,x_4,x_6) - \frac{K\left(x_3 - \frac{x_2}{NT}\right)\left(\frac{A_5}{A_3} - \left(\frac{3NT+1}{NT}\right)\cos x_3\right)}{A_4 - A_3\cos^2 x_3} \\
&\quad - \frac{A_5x_4^2\sin x_3 + A_3\sin x_3\cos x_3(x_4+x_6)^2}{A_4 - A_3\cos^2 x_3},
\end{aligned}$$

and the constants are

$$A_2 = JR_p + m_p l_2^2$$
$$A_3 = m_p l_1 l_2$$
$$A_4 = (m_1 + m_p) l_1 l_2$$
$$A_5 = (m_1 + m_p) l_1^2.$$

This DAE model is nonlinear index three and cannot be directly used in our algorithm.

Index reduction always involves some type of differentiation. For this example we start with (19) and differentiate (19g) twice replacing the derivatives using other equations. For later reference we denote the original constraint as H and the two constraints arising from differentiating twice as K, and L.

$$x_1' = x_4 \tag{20a}$$
$$x_2' = x_5 \tag{20b}$$
$$x_3' = x_6 \tag{20c}$$

$$\begin{aligned}
x_4' &= f_4(x_2, x_3, x_4, x_6) + g_{41}(x_3)u_1 - g_{41}(x_3)u_2 + \mu_1 && \text{(20d)}\\
x_5' &= f_5(x_2, x_3, x_4, x_6) - g_{41}(x_3)u_1 + g_{52}(x_3)u_2 + \mu_2 && \text{(20e)}\\
x_6' &= f_6(x_2, x_3, x_4, x_6) + g_{61}(x_3)u_1 - g_{61}(x_3)u_2 + \mu_3 && \text{(20f)}\\
0 &= \cos x_1 + \cos(x_1 + x_3) - 1 = H(x_1, x_3) && \text{(20g)}\\
0 &= -\sin(x_1)x_4 - \sin(x_1 + x_3)(x_4 + x_6) = K(x_1, x_3, x_4, x_6) && \text{(20h)}\\
0 &= -\cos(x_1)x_4^2 - \sin(x_1)(f_4 + g_{41}u_1 - g_{41}u_2 + \mu_1)\\
&\quad - \cos(x_1 + x_3)(x_4 + x_6)^2\\
&\quad - \sin(x_1 + x_3)(f_4 + g_{41}u_1 + \mu_1 - g_{41}u_2 + f_6 + g_{61}u_1 - g_{61}u_2 + \mu_3)\\
&= L(x_1, x_2, x_3, x_4, x_6, u_1, u_2, \mu_1, \mu_3) && \text{(20i)}
\end{aligned}$$

In order to make reading of the equations easier and reduce the number of subscripts we shall refer to the second model with an overline. We take the output equations as $y = C_0 x + \nu$ and $\bar{y} = C_1 \bar{x} + \bar{\nu}$. Since in the optimization problem ν, $-\bar{\nu}$ appear only as $\nu - \bar{\nu}$ we can replace them by a single variable ν which has twice as much effect. Here we take the output coefficients as

$$C_0 = \begin{bmatrix} 1 & 0 & 0 & 0 & 0 & 1 \\ 0 & 0 & 1 & 0 & 0 & 0 \end{bmatrix}, C_1 = \begin{bmatrix} 1 & 0 & 0 & 0 & 0 & 0 \\ 0 & 0 & 1 & 0 & 0 & 1 \end{bmatrix},$$

Thus the faulted model also has a change in the output.

Note (20) is an index one DAE with dynamic variables x_2, x_4, x_5, x_6 and algebraic variables x_1, x_3, u_1, u_2. However (20) is overdetermind and poses challenges for the numerical methods. In the next section we address this issue.

5.1. Solving the Nonlinear Problem

We need to reduce the number of constraining equations and make the necessary conditions index one so we return to (20). If we just drop some of the redundant algebraic constraints we get extra dynamics which might be unstable. Rather than just dropping some equations we shall use stabilized differentiation [3]. Let σ_1 and σ_2 be two negative numbers and let $(r - \sigma_1)(r - \sigma_2) = r^2 + ar + b$.

Then we take the models as

$$
\begin{aligned}
x_1' &= x_4 && \text{(21a)}\\
x_2' &= x_5 && \text{(21b)}\\
x_3' &= x_6 && \text{(21c)}\\
x_4' &= f_4(x_2,x_3,x_4,x_6) + g_{41}(x_3)u_1 - g_{41}(x_3)u_2 + \mu_1 && \text{(21d)}\\
x_5' &= f_5(x_2,x_3,x_4,x_6) - g_{41}(x_3)u_1 + g_{52}(x_3)u_2 + \mu_2 && \text{(21e)}\\
x_6' &= f_6(x_2,x_3,x_4,x_6) + g_{61}(x_3)u_1 - g_{61}(x_3)u_2 + \mu_3 && \text{(21f)}\\
L + aK + bH &= M(x_1,x_2,x_3,x_4,x_6,u_1,u_2,\mu_1,\mu_2,\mu_3) = 0 && \text{(21g)}
\end{aligned}
$$

For the inner minimization u_1, u_2 are treated as known, The x_i are differential variables and the μ_i are algebraic variables.

Then the necessary conditions for the inner problem are derived as follows. The Lagrangian for the inner minimization is

$$
\begin{aligned}
\mathcal{L} = {} & \frac{1}{2}\left(\|\mu\|^2 + \|\bar{\mu}\|^2 + 4\|\nu\|^2\right) + \lambda_1 x_4 + \lambda_2 x_5 + \lambda_3 x_6\\
& +\lambda_4\left(f_4(x_2,x_3,x_4,x_6) + g_{41}(x_3)u_1 - g_{41}(x_3)u_2 + \mu_1\right) && \text{(22a)}\\
& +\lambda_5\left(f_5(x_2,x_3,x_4,x_6) - g_{41}(x_3)u_1 + g_{52}(x_3)u_2 + \mu_2\right) && \text{(22b)}\\
& +\lambda_6\left(f_6(x_2,x_3,x_4,x_6) + g_{61}(x_3)u_1 - g_{61}(x_3)u_2 + \mu_3\right) && \text{(22c)}\\
& +\lambda_7 M + \bar{\lambda}_1\bar{x}_4 + \bar{\lambda}_2\bar{x}_5 + \bar{\lambda}_3\bar{x}_6\\
& +\bar{\lambda}_4\left(\bar{f}_4(\bar{x}_2,\bar{x}_3,\bar{x}_4,\bar{x}_6) + \bar{g}_{41}(\bar{x}_3)u_1 - \bar{g}_{41}(\bar{x}_3)u_2 + \bar{\mu}_1\right) && \text{(22d)}\\
& +\bar{\lambda}_5\left(\bar{f}_5(\bar{x}_2,\bar{x}_3,\bar{x}_4,\bar{x}_6) - \bar{g}_{41}(\bar{x}_3)u_1 + \bar{g}_{52}(\bar{x}_3)u_2 + \bar{\mu}_2\right) && \text{(22e)}\\
& +\bar{\lambda}_6\left(\bar{f}_6(\bar{x}_2,\bar{x}_3,\bar{x}_4,\bar{x}_6) + \bar{g}_{61}(\bar{x}_3)u_1 - \bar{g}_{61}(\bar{x}_3)u_2 + \bar{\mu}_3\right) && \text{(22f)}\\
& +\bar{\lambda}_7\bar{M} && \text{(22g)}\\
& +\lambda_8\left(x_1 + x_6 - \bar{x}_1 + \nu_1\right) + \bar{\lambda}_8\left(x_3 - \bar{x}_3 - \bar{x}_6 + \nu_2\right) && \text{(22h)}
\end{aligned}
$$

The terms in (22h) are from the additional constraint in the inner minimization problem that $y = \bar{y}$.

This leads to the outer minimization problem which is now

$$\min \int_0^T \|u_1\|^2 + \|u_2\|^2 dt \tag{23}$$

subject to

$$
\begin{aligned}
x_1' &= x_4 && \text{(24a)}\\
x_2' &= x_5 && \text{(24b)}\\
x_3' &= x_6 && \text{(24c)}\\
x_4' &= f_4(x_2,x_3,x_4,x_6) + g_{41}(x_3)u_1 - g_{41}(x_3)u_2 + \mu_1 && \text{(24d)}\\
x_5' &= f_5(x_2,x_3,x_4,x_6) - g_{41}(x_3)u_1 + g_{52}(x_3)u_2 + \mu_2 && \text{(24e)}\\
x_6' &= f_6(x_2,x_3,x_4,x_6) + g_{61}(x_3)u_1 - g_{61}(x_3)u_2 + \mu_3 && \text{(24f)}\\
0 &= M(x_1,x_2,x_3,x_4,x_6,u_1,u_2,\mu_1,\mu_2,\mu_3) = 0 && \text{(24g)}\\
\bar{x}_1' &= \bar{x}_4 && \text{(24h)}\\
\bar{x}_2' &= \bar{x}_5 && \text{(24i)}\\
\bar{x}_3' &= \bar{x}_6 && \text{(24j)}\\
\bar{x}_4' &= \bar{f}_4(\bar{x}_2,\bar{x}_3,\bar{x}_4,\bar{x}_6) + \bar{g}_{41}(\bar{x}_3)u_1 - \bar{g}_{41}(\bar{x}_3)u_2 + \bar{\mu}_1 && \text{(24k)}\\
\bar{x}_5' &= \bar{f}_5(\bar{x}_2,\bar{x}_3,\bar{x}_4,\bar{x}_6) - \bar{g}_{41}(\bar{x}_3)u_1 + \bar{g}_{52}(\bar{x}_3)u_2 + \bar{\mu}_2 && \text{(24l)}\\
\bar{x}_6' &= \bar{f}_6(\bar{x}_2,\bar{x}_3,\bar{x}_4,\bar{x}_6) + \bar{g}_{61}(\bar{x}_3)u_1 - \bar{g}_{61}(\bar{x}_3)u_2 + \bar{\mu}_3 && \text{(24m)}\\
0 &= \bar{M}(\bar{x}_1,\bar{x}_2,\bar{x}_3,\bar{x}_4,\bar{x}_6,u_1,u_2,\bar{\mu}_1,\bar{\mu}_2,\bar{\mu}_3) = 0 && \text{(24n)}\\
0 &= x_1 + x_6 - \bar{x}_1 + \nu_1 && \text{(24o)}\\
0 &= x_3 - \bar{x}_3 - \bar{x}_6 + \nu_2 && \text{(24p)}
\end{aligned}
$$

$$
\begin{aligned}
-\lambda_1' &= M_{x_1}\lambda_7 + \lambda_8 && \text{(25a)}\\
-\lambda_2' &= f_{4,x_2}\lambda_4 + f_{5,x_2}\lambda_5 + f_{6,x_2}\lambda_6 + M_{x_2}\lambda_7 && \text{(25b)}\\
-\lambda_3' &= (f_{4,x_3} + g_{41,x_3}u_1 - g_{41,x_3}u_2)\lambda_4\\
&\quad (f_{5,x_3} - g_{41,x_3}u_1 + g_{52,x_3}u_2)\lambda_5\\
&\quad (f_{6,x_3} + g_{61,x_3}u_1 - g_{61,x_3}u_2)\lambda_6 + M_{x_3}\lambda_7 + \bar{\lambda}_8 && \text{(25c)}\\
-\lambda_4' &= \lambda_1 + f_{4,x_4}\lambda_4 + f_{5,x_4}\lambda_5 + f_{6,x_4}\lambda_6 + M_{x_4}\lambda_7 && \text{(25d)}\\
-\lambda_5' &= \lambda_2 && \text{(25e)}\\
-\lambda_6' &= \lambda_3 + f_{4,x_6}\lambda_4 + f_{5,x_6}\lambda_5 + f_{6,x_6}\lambda_6 + M_{x_6}\lambda_7 + \lambda_8 && \text{(25f)}\\
-\bar{\lambda}_1' &= \bar{M}_{\bar{x}_1}\bar{\lambda}_7 - \lambda_8 && \text{(25g)}\\
-\bar{\lambda}_2' &= \bar{f}_{4,\bar{x}_2}\bar{\lambda}_4 + \bar{f}_{5,\bar{x}_2}\bar{\lambda}_5 + \bar{f}_{6,\bar{x}_2}\bar{\lambda}_6 + \bar{M}_{\bar{x}_2}\bar{\lambda}_7 && \text{(25h)}\\
-\bar{\lambda}_3' &= (\bar{f}_{4,\bar{x}_3} + \bar{g}_{41,\bar{x}_3}u_1 - \bar{g}_{41,\bar{x}_3}u_2)\bar{\lambda}_4\\
&\quad (\bar{f}_{5,\bar{x}_3} - \bar{g}_{41,\bar{x}_3}u_1 + \bar{g}_{52,\bar{x}_3}u_2)\bar{\lambda}_5\\
&\quad (\bar{f}_{6,\bar{x}_3} + \bar{g}_{61,\bar{x}_3}u_1 - \bar{g}_{61,\bar{x}_3}u_2)\bar{\lambda}_6 + M_{x_3}\bar{\lambda}_7 - \bar{\lambda}_8 && \text{(25i)}\\
-\bar{\lambda}_4' &= \bar{\lambda}_1 + \bar{f}_{4,\bar{x}_4}\bar{\lambda}_4 + \bar{f}_{5,\bar{x}_4}\bar{\lambda}_5 + \bar{f}_{6,\bar{x}_4}\bar{\lambda}_6 + \bar{M}_{\bar{x}_4}\bar{\lambda}_7 && \text{(25j)}\\
-\bar{\lambda}_5' &= \bar{\lambda}_2\\
-\bar{\lambda}_6' &= \bar{\lambda}_3 + \bar{f}_{4,\bar{x}_6}\bar{\lambda}_4 + \bar{f}_{5,\bar{x}_6}\bar{\lambda}_5 + \bar{f}_{6,\bar{x}_6}\bar{\lambda}_6 + \bar{M}_{\bar{x}_6}\bar{\lambda}_7 - \bar{\lambda}_8 && \text{(25k)}
\end{aligned}
$$

and

$$\begin{aligned}
0 &= \mu_1 + \lambda_4 + M_{\mu_1}\lambda_7 && \text{(26a)}\\
0 &= \mu_2 + \lambda_5 + M_{\mu_2}\lambda_7 && \text{(26b)}\\
0 &= \mu_3 + \lambda_6 + M_{\mu_3}\lambda_7 && \text{(26c)}\\
0 &= \bar{\mu}_1 + \bar{\lambda}_4 + \bar{M}_{\bar{\mu}_1}\bar{\lambda}_7 && \text{(26d)}\\
0 &= \bar{\mu}_2 + \bar{\lambda}_5 + \bar{M}_{\bar{\mu}_2}\bar{\lambda}_7 && \text{(26e)}\\
0 &= \bar{\mu}_3 + \bar{\lambda}_6 + \bar{M}_{\bar{\mu}_3}\bar{\lambda}_7 && \text{(26f)}\\
0 &= 4\nu + \lambda_8 && \text{(26g)}\\
0 &= 4\bar{\nu} + \bar{\lambda}_8 && \text{(26h)}\\
\psi' &= \frac{1}{2}\left(\|\mu\|^2 + \|\bar{\mu}\|^2 + 4\|\nu\|^2\right) && \text{(26i)}\\
\psi(T) &\geq \gamma^2, && \text{(26j)}
\end{aligned}$$

along with the boundary conditions

$$\begin{aligned}
\psi(0) &= 0 && \text{(27a)}\\
\lambda_i(T) &= 0 && \text{(27b)}\\
\bar{\lambda}_i(T) &= 0 && \text{(27c)}\\
x(0) &= x_s && \text{(27d)}\\
\bar{x}(0) &= x_s && \text{(27e)}
\end{aligned}$$

The set points used in the linearized models correspond to the arm holding a mass motionless. Calculation of a set point will determine part, but not all, of u. Let u_1 be the free component. Thus, it is an algebraic variable in terms of the DAE, while u_2 is the test signal. The set points for the nominal and faulty models are found via a nonlinear solve in Matlab. The nominal model set point is found using a relatively large value for m_p to simulate an object being held. We will analyze one faults and refer to the faulty model as model 1. Model 1, simulates a breakdown of the second joint by lowering the elasticity parameter K. Model 0 is the nominal model. The set point is given in Table 1 and values of all parameters are given in Table 2.

For out illustration we take $T = 10$. The value of K is decreased from 4 to 1 in the faulty model. Thus the fault consists of both a change in K and also a change in the sensor output matrix C.

The results are given in Figures 2–4. Figure 2 gives the proper test signal.

Table 1. Numerically computed set point with $m_p = 10$

x_1	.565
x_2	1.70
x_3	.85
x_4	0
x_5	0
x_6	0
u_1	0
u_2	0

Table 2. Parameter values for the robot arm

Parameter	**Description**	**Nominal Value**
K	Coefficient of elasticity at joint 2	$K_0 = 4$, $K_1 = 1$
NT	Transmission ratio of joint 2	2
JR_0	Rotor inertia of joint 1	1
JR_1	Rotor inertia of joint 2	1
JR_p	Inertia at endpoint	1
l_1	Length of link 1	1
l_2	Length of link 2	1
m_1	Mass of joint 2	1
m_p	Mass at endpoint	10

Theoretically we should have H and L zero. The values of $H, \bar{H}, L, \bar{L}$ were monitored in all of the computational tests. The graphs of H and $\bar{H}$ and also $L, \bar{L}$ were always very similar so we will usually only give one of each pair. Figure 3 shows H and L. Our x_s actually gives a value of H of 10^{-4} and L of 0. As expected H goes to zero exponentially fast and L rises briefly and then goes to zero exponentially fast. The graph of ψ is given in Figure 4. The ψ graph serves as a check that the code is computing correctly. ψ should start at 0 for our scenario and rise monotonically until it reaches γ^2 at time T.

The use of stabilized differentiation introduces extra dynamics in the mod-

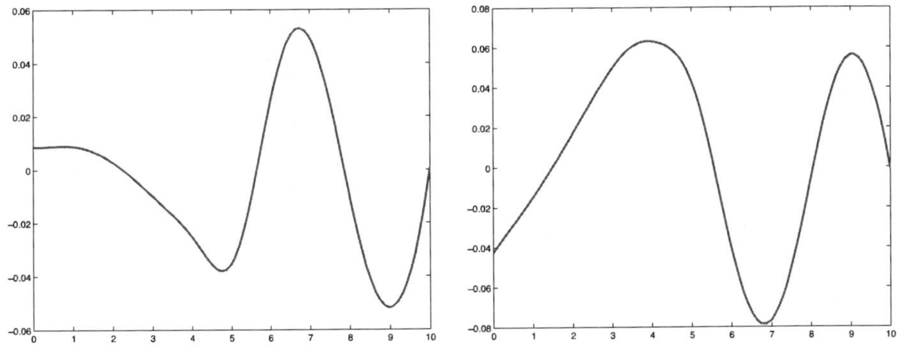

Figure 2. Test signals computed from (23)–(27) using $\sigma = -1$, $\sigma = -2$, $\gamma^2 = 2$.

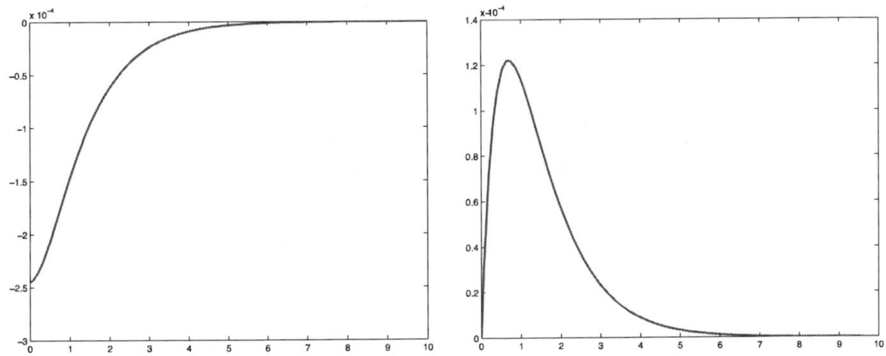

Figure 3. Error in H (left) and L (right) for (23)–(27) with $\sigma = -1$, $\sigma = -2$ $\gamma^2 = 2$.

els. To see the effect of these extra dynamics we rerun this example but with σ values of -4 and -5 rather than -1 and -2 so that there is faster stabilization. The results are shown in Figures 5–7. The new test signals are shown in Figure 5 and should be compared to Figure 2. We see that the test signal is of similar size but somewhat different shape. The only difference in the two cases is that the extra dynamics are different so the computed test signals are minimum proper for slightly different systems. However, all the dynamics of the original prob-

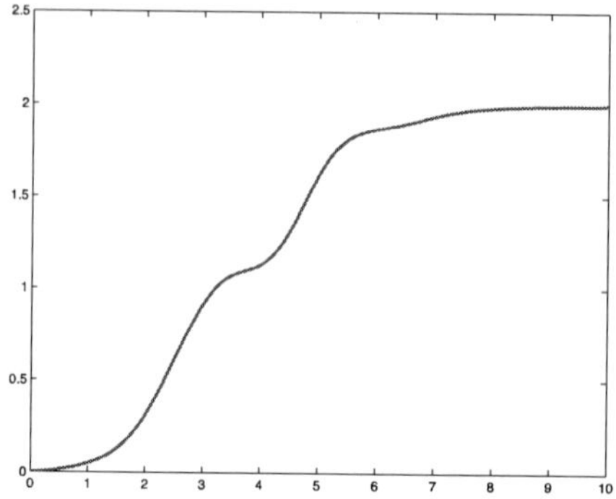

Figure 4. Graph of ψ computed from (23)–(27) with $\sigma = -1$, $\sigma = -2$, $\gamma^2 = 2$.

lem are included in the dynamics of any stabilized system. Thus the test signals of Figures 2 and 5 will be proper for the system of interest and will be close to minimum proper and thus good test signals. In Figure 6 we have graphed the error in H and $\bar{H}$ to show that they are in fact essentially identical.

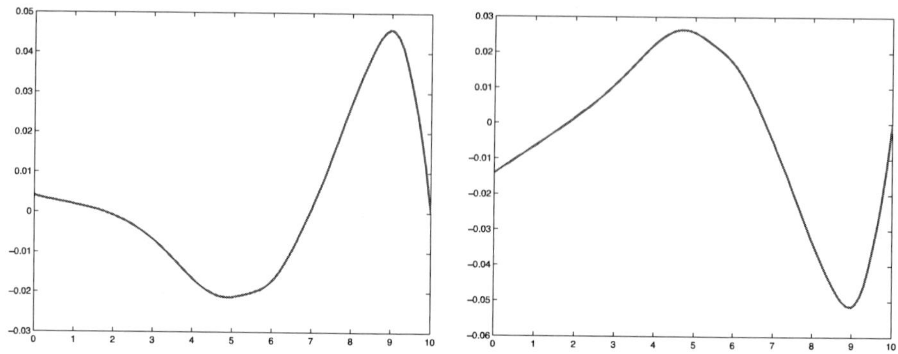

Figure 5. Test signals computed from (23)–(27) with $\sigma = -4$, $\sigma = -4$, $\gamma^2 = 2$.

Fault detection in nonlinear ODE models using linearizations was studied in [36]. In the ODE case, the author shows auxiliary signals can sometimes be found for the nonlinear problem by linearizing the nonlinear models and finding a minimal proper signal for the linearized models. The work in [36] shows if u is strictly proper for the linearized system and β is small enough and a positive

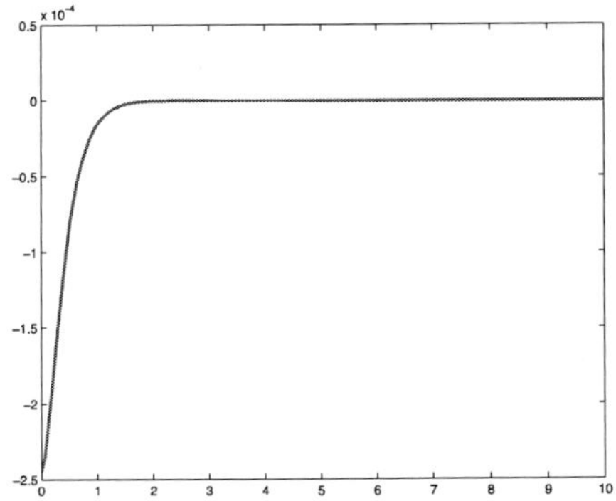

Figure 6. Error in H (left) and $\bar{H}$ (right) from (23)–(27) with $\sigma = -4$, $\sigma = -5$, $\gamma^2 = 2$.

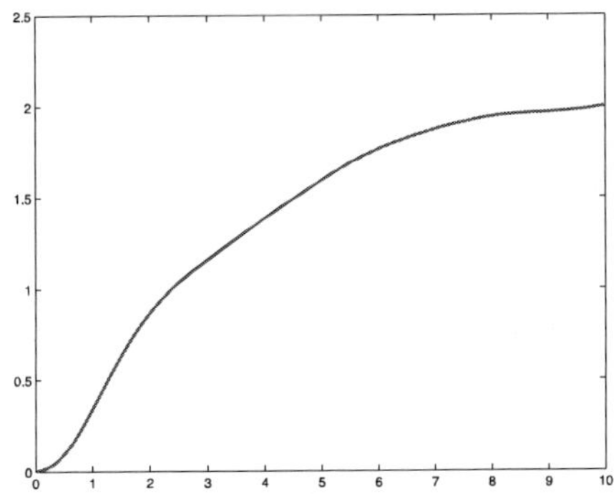

Figure 7. Graph of ψ from (23)–(27) with $\sigma = -4$, $\sigma = -5$, $\gamma^2 = 2$.

scalar, then βu is proper for the nonlinear system with the noise bounds scaled by β^2. Furthermore, if the nonlinearities satisfy certain technical assumptions, then u is proper for the nonlinear problem, except for possibly tighter noise bounds.

These results from the literature all assumed that one could operate close enough to the set point. Depending on the strength of the nonlinearity that might be very restrictive on the amount of uncertainty that could be permitted. A major advantage of the nonlinear approach of this paper is that we can get proper test signals in the presence of more uncertainty. To illustrate we reran the preceding example except that now we set $\gamma^2 = 10$ which allows for much more uncertainty. The result is in Figures 8 and 9. Comparing Figure 8 to Figures 2 and 5 we see that not only is the test signal larger as expected but it has a very different shape which means the signal is not just a scaled version of the smaller uncertainty test signal. The test signal found from a linearization might well not be acceptable with γ this large.

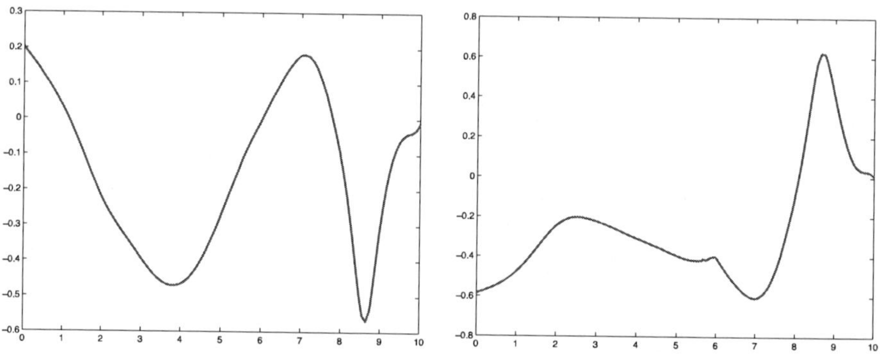

Figure 8. Test signals computed from (23)–(27) with $\sigma = -4$, $\sigma = -5$, $\gamma^2 = 10$.

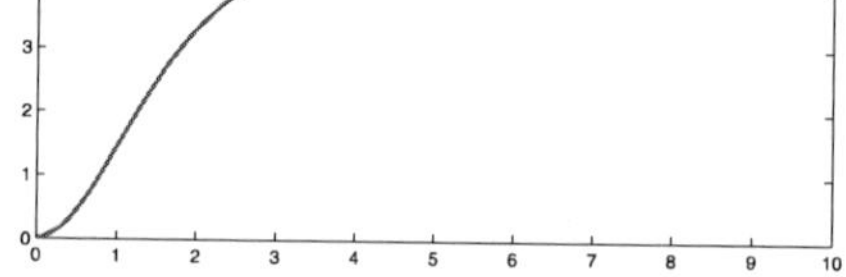

Figure 9. Graph of ψ computed from (23)–(27) with $\sigma_1 = -4$, $\sigma_2 = -5$, $\gamma^2 = 10$.

Conclusion

In this paper we have examined the application of a previous fault detection technique that was initially developed for linear DAE models, to a nonlinear DAE robot arm model.

First a general formulation was presented and discussed. For some applications resulting in higher index DAEs this initial formulation is not in a good form for computation. In reformulating the models if they are higher index, then some type of differentiation is needed. However, it is desirable to only differentiate known equations and not differentiate noise or computed quantities. We present two possible approaches.

In the general discussion constraints are added and the number of dynamic equations is reduced to give index one systems. Our experience is solving a large number of optimal control problems with DAE models and direct transcription software is that having lots of additional constraints can create difficulty in getting feasible solutions on the initial coarse grids. Therefore we illustrate an alternative on the computational case study. We use stabilized differentiation and keep all of the dynamics. Mathematically this means that some of the uncertainty becomes algebraic state variables rather than "controls" on the inner minimization problem. The stabilized dynamics index one systems have more solutions in its behavior than the original DAE model. But any test signal proper for the stabilized system will also be proper for the original DAE.

It was seen that the nonlinear approach could compute good test signals even when there were higher levels of uncertainty and linearization would not be appropriate. It was also seen that the use of stabilized differentiation could produce good test signals for the original problem that were close to minimal although the stabilization parameters could affect the shape of the test signal.

Acknowledgment

Research supported in part by NSF Grant DMS-1209251.

References

[1] I. Andjelkovic, K. A. Sweetingham, and S. L. Campbell. Active fault detection in nonlinear systems using auxiliary signals. In Proc. American Control Conference, Seattle, WA, 2008.

[2] C. M. Astorga-Zaragoza, D. Theilliol, J. C. Ponsart, and M. Rodrigues. Fault diagnosis for a class of descriptor linear parameter varying systems. *Int. Journal of Adaptive Control and Signal Processing*, 26(3):208–223, 2012.

[3] J. Baumgarte. Stabilization of constraints and integrals of motion in dynamical systems. *Comp. Meth. Appl. Mech. Eng.*, 1:1–16, 1972.

[4] G. Besanćon, I. Rubio-Scola, and D. Georges. Input selection in observer design for non-uniformly observable systems. In 9th IFAC Symposium on Nonlinear Control Systems, 2013.

[5] J. T. Betts. *Practical Methods for Optimal Control and Estimation using Nonlinear Programming*. SIAM, 2nd ed. edition, 2010.

[6] K. Brenan, S. L. Campbell, and L. R. Petzold. *Numerical Solution of Initial Value Problems in Differential-Algebraic Equations*. SIAM, Philadelphia, PA, 1996.

[7] S. L. Campbell and R. Marz. Direct transcription solution of high index optimal control problems and regular euler-lagrange equations. *J. Comp. Appl. Math.*, 202:186–202, 2007.

[8] S. L. Campbell and R. Nikoukhah. *Auxiliary Signal Design for Failure Detection*. Princeton University Press, Princeton, New Jersey, 2004.

[9] S. L. Campbell and J. Scott. Active fault detection in nonlinear differential algebraic equations II: Using linearizations. Preprint, 2016.

[10] D. Choe, S. L. Campbell, and R. Nikoukhah. A comparison of optimal and suboptimal auxiliary signal design approaches for robust failure detection. In Proc. IEEE Conference on Control Applications, Toronto, 2005.

[11] M. L. Corradini, A. Cristofaro, and S. Pettinari. Robust FDI filters and fault sensitivity analysis in continuous-time descriptor systems. In Proc. 51st IEEE Conference on Decision and Control, pages 1220–1225, Maui, HI, December 2012.

[12] M. Darouach and M. Boutayeb. Design of observers for descriptor systems. *IEEE Transactions on Automatic Control*, 40:1323–1327, 1995.

[13] A. De Luca and A. Isidori. Feedback linearization of invertible systems. In 2nd Duisburger Kolloquium Automation und Robotik, Duisburg, Germany, July 1987.

[14] De Luca, A. Control properties of robot arms with joint elasticity. In Proc. International Symposium on the Mathematical Theory of Networks and Systems, Phoenix, AZ, June 1987.

[15] K. J. Drake, S. L. Campbell, I. Andjelkovic, and K. Sweetingham. Active incipient failure detection: a nonlinear case study. In Proc. 4th International Conference on Computing, Communications and Control Technologies, Orlando, 2006.

[16] G. R. Duan, D. Howe, and R. J. Patton. Robust fault detection in descriptor linear systems via generalized unknown input observers. *International Journal of Systems Science*, 33(5):369–377, 2010.

[17] A. Engelsone, S. L. Campbell, and J. T. Betts. Direct transcription solution of higher-index optimal control problems and the virtual index. *Applied Numerical Mathematics*, 57:281–296, 2007.

[18] Z. Gao and H. Wang. Descriptor observer approaches for multivariable systems with measurement noises and application in fault detection and diagnosis. *Systems and Control Letters*, 55(4):304–313, 2006.

[19] M. Gerdin, T. Glad, and L. Ljung. Parameter estimation in linear differential-algebraic equations. In Proc. 13th IFAC Symposium on System Identification, 2003.

[20] M. Gerdts. Parameter identification in higher index DAE systems. Technical report, Department of Mathematics, Universität Hamburg, 2005.

[21] P. E. Gill, W. Murray, and M. A. Saunders. User's guide for SNOPT 7: A Fortran package for large-scale nonlinear programming. Systems Optimization Laboratory, Stanford University, CA 94305-4023, 2007.

[22] P. Kunkel and S. L. Campbell. Solving higher index DAE optimal control problems. Preprint, 2016.

[23] P. Kunkel and V. Mehrmann. *Differential-Algebraic Equations. Analysis and Numerical Solution*. EMS Publishing House, Zürich, Switzerland, 2006.

[24] DJ. Lee, Y. Park, and YS Park. Robust H_∞ sliding mode descriptor observer for fault and output disturbance estimation of uncertain systems. *IEEE Transactions on Automatic Control*, 57(11):2928–2934, 2012.

[25] H. H. Niemann. Active fault diagnosis in closed-loop uncertain systems. In *6th IFAC Symposium on Fault Detection Supervision and Safety for Technical Processes*, pages 587–592, Beijing, China, 2006.

[26] H. H. Niemann. A setup for active fault diagnosis. *IEEE Transactions on Automatic Control*, 51(9):1572–1578, 2006.

[27] N. Poulsen and H. H. Niemann. Active fault diagnosis based on stochastic tests. *International Journal of Applied Mathematics and Computer Science*, 18:487–496, 2008.

[28] N. K. Poulsen and H. H. Niemann. Active fault diagnosis–a stochastic approach. In 7th IFAC Symposium on Fault Detection Supervision and Safety for Technical Processes, 2009.

[29] A. V. Rao, D. A. Benson, C. Darby, M. A. Patterson, C. Francolin, I. Sanders, and G. T. Huntington. Algorithm 902: Gpops, a matlab software for solving multiple-phase optimal control problems using the Gauss pseudospectral method. *ACM Transactions on Mathematical Software*, 37(37):22:1–22:39, 2010.

[30] I. Rubio-Scola, G. Besanćon, and D. Georges. Input optimization for observability of state affine systems. In Proc. 5th IFAC Symposium on System Structure and Control, 2013.

[31] I. Rubio-Scola, G. Besanćon, and D. Georges. Online observability optimization for state affine systems with output injection and observer design. In Proc. 21st IEEE Mediterranean Conference on Control and Automation, 2013.

[32] J. Scott and S. L. Campbell. Observer based fault detection and identification in differential algebraic equations. In Proc. ASME 6th Annual Dynamic Systems and Control Conference, page V002T24A001; 10 pages, Stanford, California, 2013.

[33] J. Scott and S. L. Campbell. Observer based fault detection in differential algebraic equations. In Proc. SIAM Control Conference, pages 176–183, San Diego, CA, 2013.

[34] J. R. Scott and S. L. Campbell. Auxiliary signal design for failure detection in differential-algebraic equations. *Numerical Algebra, Control and Optimization*, 4:151–179, 2014.

[35] J. R. Scott and S. L. Campbell. Auxiliary signal design for failure detection in high index differential-algebraic equations. In Proc. 2014 IEEE Conference on Decision and Control, Los Angeles, CA, 2014.

[36] K. Sweetingham and S. Campbell. Auxiliary signal design for fault detection in nonlinear systems. *Control and Intelligent Systems*, 37:152–159, 2009.

[37] A. Varga. *Descriptor system techniques in solving $\mathcal{H}_{2/\infty}$-optimal fault detection and isolation problems*, volume 23 of *Advances in Design and Control*, Chap. 6, pages 107–128. SIAM, 2012.

[38] A. Wächter and L. T. Biegler. On the implementation of a primal duel interior point filter line search algorithm for large-scale nonlinear programming. *Mathematical Programming*, 106(1):25–57, 2006.

[39] T. K. Yeu, H. S. Kim, and S. Kawaji. Fault detection, isolation and reconstruction for descriptor systems. *Asian Journal of Control*, 7(4):356–367, 2005.

In: Fault Detection
Editor: Daniel Martin
ISBN: 978-1-53610-345-8
© 2017 Nova Science Publishers, Inc.

Chapter 2

ACTIVE FAULT DETECTION IN NONLINEAR DIFFERENTIAL ALGEBRAIC EQUATIONS II: USING LINEARIZATIONS

Stephen L. Campbell [*] ***and Jason R. Scott*** [†]
Department of Mathematics, North Carolina State University,
Raleigh, North Carolina, US

Abstract

Fault detection and identification (FDI) play a fundamental role in most modern industrial systems and processes. They are essential in ensuring safe and reliable operation. A variety of approaches, both active and passive have been investigated. Most of that work required the models to be linear and modeled by ordinary differential equations. Recently, new results on auxiliary signal design for fault detection in linear differential algebraic equations (DAEs) has led to a technique that can potentially be used on some linear DAE models. This paper examines this application with a nontrivial case study involving a nonlinear robot arm model. It is shown the previous theory does not immediately apply to the linearizations and some modifications are sometimes necessary in order to use this approach on the linearized models that occur in some applications. Techniques for reformulating are presented. It is also shown how to evaluate a test signal for a nonlinear problem found by this or any other method.

[*] E-mail address: slc@ncsu.edu
[†] E-mail address: jrscott@ncsu.edu

Keywords: Fault detection, differential algebraic equations, linearization

AMS Subject Classification: 53D, 37C, 65P

1. Introduction

In a multi-model active approach for fault detection we assume there are known models for the nominal and faulty behaviors of the system. An auxiliary signal is applied whose purpose is to facilitate detection by forcing the input-output sets for the nominal and faulty models to be disjoint. An additional goal in auxiliary signal design is to find the smallest such signal that guarantees detection, thereby limiting the effect of the signal on the system. Active detection has been studied extensively for ODE systems in [4, 11, 12]. Linear differential algebraic equation (DAE) models are considered in [15].

The companion paper [2] examines working directly with nonlinear DAE models. This paper differs from [2] in several ways. This paper considers the use of linearizations. A somewhat different fault scenario is examined. This paper also shows how to take the test signal from the linearization and evaluate how it performs on the original nonlinear problem.

Additional related references on fault detection and DAEs may be found in [2]. However, that work was generally limited to linear and index one DAEs. The approach of [15] can be used for some higher index linear DAE models. However, [15] requires some technical assumptions and cannot always be used immediately on a DAE model arising in an application. In this paper we will use an example from robotics to illustrate some of the modifications that need to be made in order for this approach to work on the higher index DAE model case.

Section 2 will quickly review the basic ideas of [15, 16]. Applications are not always initially formulated so that they immediately meet the assumptions of [15]. Section 3 carefully discusses several issues in the context of a nonlinear control problem. For one, it illustrates how to get a reformulation so that the techniques of [15] can be used. Different reformulation options can be used. The one sought here is different from that sought in [2]. In Section 4 it is shown how to verify a candidate test signal, however it is found, is proper. This is important both for test signals found from linearizations as in this paper that will be used on nonlinear systems and also on test signals found or designed by other methods such as that of [2]. Finally Section 5 covers using the test signal on a nonlinear system Conclusions are in Section 6.

We assume that the reader is familiar with the basic theory of DAEs and their index [1, 8].

2. Review of Linear DAE Approach

We quickly review some ideas from [15]. Consider the following multi-model formulation with two models and additive uncertainty. One model is the nominal system ($i = 0$) and the other is the faulty system ($i = 1$). Each model has the form

$$E_i x_i' + F_i x_i = B_i u + M_i \mu_i \tag{1a}$$

$$y_i = C_i x_i + D_i u + N_i \eta_i. \tag{1b}$$

Here u is the test signal to be applied. Matrices E_i, F_i, B_i, M_i, C_i, D_i, N_i are constant in the original models. E_i may be singular allowing, in general, the models to be DAEs. y_0, y_1 must have the same dimension, but x_0, x_1 do not have to have the same dimension, allowing for the consideration of structural faults. The ν_i, η_i are additive uncertainties representing noise or model error. There are several variations of this problem such as including known exogenous input and feedthrough terms that can be easily handled given the information provided in this paper. The N_i are assumed nonsingular. That is, we allow noise or uncertainty in all equations. The test is to be conducted on the interval $0 \le t \le T$.

For each model the uncertainty consists of $x_i(0)$, μ_i, ν_i and is measured by

$$\Gamma_i^2(x_i(0), \mu_i, \nu_i) \quad = \quad \frac{1}{2} x_i(0)^T P_{0_i} x_i(0) + \frac{1}{2} \int_0^T \|\nu_i\|^2 + \|\mu_i\|^2 \, dt. \tag{2}$$

Two ways of measuring the total uncertainty in both models have been considered. We will use

$$\Gamma_2^2 = \Gamma_0^2 + \Gamma_1^2 \tag{3}$$

which was introduced in [5]. We will assume that the total amount of uncertainty is bounded by γ so that $\Gamma_2^2 \le \gamma^2$.

A test signal u is proper if it is not possible to get the same output from both models and still have the total uncertainty bound hold. That is, the smallest amount of uncertainty for which $y_1 = y_0$ violates the total uncertainty bound γ^2. We seek the smallest proper test signal. There are several ways to measure the size of the test signal [4]. In this paper we shall use the energy or L^2 bound,

$$\|u\|^2 = \int_0^T \|u(t)\|^2 dt. \tag{4}$$

Use of a different measure of u is usually easy to implement since it amounts to altering one line of code where the outer cost is given. Several examples are given in [4].

Since we want the minimal proper signal, we must solve

Problem 1.

$$\textit{Find } J = \inf_u ||u||^2 \tag{5a}$$

such that

$$\left(\inf_{\substack{x_i(0),\mu_i,\eta_i \\ i=0,1}} \Gamma^2(u) \right) \geq \gamma^2 \tag{5b}$$

$$E_0 x_0' + F_0 x_0 = B_0 u + M_0 \mu_0 \tag{5c}$$

$$E_1 x_1' + F_1 x_1 = B_1 u + M_1 \mu_1 \tag{5d}$$

$$0 = y_0 - y_1. \tag{5e}$$

The solution of Problem 1 is known as the minimal proper u with respect to the L^2 bound. Frequently, the minimal proper u is unique up to a factor of negative one. However, examples can be constructed where it is not unique. From [15], a reformulation of the problem, making it easier to solve numerically is gotten by replacing (5b) with the necessary conditions for this minimization and also adding a collocation to compute the cost of the inner minimization.

Fully describing how Problem 1 can be reformulated and explicitly giving all the transformations would run for several pages. Here we just outline the process for the reader's benefit. Full details can be found in [15].

Necessary conditions of the inner minimum in (5) can be obtained by performing numerically robust linear time invariant coordinate changes on the system (5c)–(5e) to get

$$
\begin{aligned}
x'_{01} &= A_{011}x_{01} + A_{012}x_{02} + B_{01}u + M_{011}\mu_{01} + M_{012}\mu_{02} && \text{(6a)}\\
0 &= A_{021}x_{01} + A_{022}x_{02} + B_{02}u + M_{022}\mu_{02} && \text{(6b)}\\
x'_{11} &= A_{111}x_{11} + A_{112}x_{12} + B_{11}u + M_{111}\mu_{11} + M_{112}\mu_{12} && \text{(6c)}\\
0 &= A_{121}x_{11} + A_{122}x_{12} + B_{12}u + M_{122}\mu_{12} && \text{(6d)}\\
0 &= C_{01}x_{01} + C_{02}x_{02} + D_0u + N_0\eta_0 \\
&\quad -(C_{11}x_{11} + C_{12}x_{12} + D_1u + N_1\eta_1). && \text{(6e)}
\end{aligned}
$$

Unlike in most of the literature, here A_{i22} are not assumed to be nonsingular so that the original DAE models need not be index one. Hence, A_{i22} may have a nontrivial null space. A_{i12} may also have a nontrivial null space. If $x_{i2} \in N(A_{i22}) \cap N(A_{i12})$, then x_{i2} does not affect the dynamics or the constraints and is a free control variable. Since this may cause problems for the optimizer, we will assume that $N(A_{i22}) \cap N(A_{i12}) = \emptyset$.

Algebraic variables include both controls and undifferentiated state variables. Let $\zeta_1 = \begin{bmatrix}\mu_{02}^T & \mu_{12}^T & \eta_0^T\end{bmatrix}^T$ and $\zeta_2 = \begin{bmatrix}\mu_{01}^T & \mu_{11}^T & \eta_1^T\end{bmatrix}^T$. Then (6) has the form

$$
z'_1 = \hat{A}z_1 + \begin{bmatrix}\hat{D} & \hat{N}\end{bmatrix}\begin{bmatrix}z_2\\ \zeta_1\end{bmatrix} + \hat{B}u + \hat{M}\zeta_2 \tag{7a}
$$

$$
0 = \tilde{A}z_1 + \begin{bmatrix}\tilde{D} & \tilde{N}\end{bmatrix}\begin{bmatrix}z_2\\ \zeta_1\end{bmatrix} + \tilde{B}u + \tilde{M}\zeta_2 \tag{7b}
$$

where

$$
\begin{aligned}
\tilde{D} &= \begin{bmatrix} A_{022} & 0 \\ 0 & A_{122} \\ C_{02} & -C_{12}\end{bmatrix},\\
z_1 &= \begin{bmatrix} x_{01}\\ x_{11}\end{bmatrix},\ z_2 = \begin{bmatrix} x_{02}\\ x_{12}\end{bmatrix},
\end{aligned} \tag{8a}
$$

and $\tilde{N}$ is invertible. Assume $\tilde{D}$ has full column rank. $\begin{bmatrix}\tilde{D} & \tilde{N}\end{bmatrix}$ has full row rank and $\tilde{N}$ is invertible so we can pick a subset of the columns of $\tilde{N}$, call them $\tilde{N}_1$, so that $\begin{bmatrix}\tilde{D} & \tilde{N}_1\end{bmatrix}$ is invertible. Let $\tilde{N}_2$ denote the remaining columns. Let $\zeta_1 = \begin{bmatrix}\zeta_{11}^T & \zeta_{12}^T\end{bmatrix}^T$ be a conformal partition. Then

$$
-\begin{bmatrix} z_2\\ \zeta_{11}\end{bmatrix} = \begin{bmatrix}\tilde{D} & H_1\end{bmatrix}^{-1}\left(\tilde{A}z_1 + \tilde{N}_2\zeta_{12} + \tilde{B}u + \tilde{M}\zeta_2\right). \tag{9}
$$

Let $\hat{R} = \begin{bmatrix} \hat{D} & \hat{N}_1 \end{bmatrix}$, where $\hat{N}_1$ are the columns of $\hat{N}$ that are conformal to the selection made for $\tilde{N}_1$. Eliminating $\begin{bmatrix} z_2 \\ \zeta_{11} \end{bmatrix}$ from the system, (7) can be rewritten as

$$z_1' = Az_1 + Bu + N\zeta_{12} + M\zeta_2, \tag{10}$$

where $A = \hat{A} - \hat{R}\begin{bmatrix} \tilde{D} & \tilde{N}_1 \end{bmatrix}^{-1}\tilde{A}$ and B, N, M are defined in the analogous way.

Due to the orthogonality of the transformations on μ_i and the coordinate changes made on x_i, the total uncertainty using the L^2 noise measure is

$$\Gamma^2 = \frac{1}{2}z^T(0)\, Pz(0) + \frac{1}{2}\int_0^T \|\zeta_{11}\|^2 + \|\zeta_{12}\|^2 + \|\zeta_2\|^2\, dt \tag{11}$$

where P is computed from the original cost and a robustification of the optimization problem by adding some additional initial uncertainty. This approach is conservative in that it allows more noise than the original problem. Hence, the proper u found with this noise bound will still be proper for the original problem.

Minimizing Γ^2 yields the same solution as minimizing Γ so the right-hand side of (11) can be minimized directly. ζ_{11} has been solved for in terms of the other variables. Let $\tilde{A}_0$ denote the bottom half of the matrix $\tilde{A}$ where the partition is conformal with $\begin{bmatrix} z_2^T & \zeta_{11}^T \end{bmatrix}$ and do the same for the other matrices in (9). Then, the inner cost takes the form

$$\Gamma^2 = \frac{1}{2}z_1^T(0)Pz_1(0) + \frac{1}{2}\int_0^T I(z_1, \zeta_{12}, \zeta_2)dt. \tag{12}$$

An explicit form for $I(z_1, \zeta_{12}, \zeta_2)$ is found in [15]. Then the Hamiltonian for this inner problem can be defined by

$$H(z_1, \zeta_{12}, \zeta_2) = \frac{1}{2}I(z_1, \zeta_{12}, \zeta_2) + \lambda^T(Az_1 + Bu + G\zeta_{12} + M\zeta_2) \tag{13}$$

where λ is the Lagrange multiplier. The dynamic portion of the necessary conditions for a minimum of Γ are

$$\begin{aligned} z_1' &= Az_1 + Bu + G\zeta_{12} + M\zeta_2 && \text{(14a)} \\ -\lambda' &= A^T\lambda + \tilde{A}_0^T\tilde{A}_0 z_1 + \tilde{A}_0^T\tilde{G}_0\zeta_{12} + \tilde{A}_0^T\tilde{B}_0 u + \tilde{A}_0^T\tilde{M}_0\zeta_2 && \text{(14b)} \\ 0 &= G^T\lambda + \tilde{G}_0^T\tilde{A}_0 z_1 + (I + \tilde{G}_0^T\tilde{G}_0)\zeta_{12} + \tilde{G}_0^T\tilde{B}_0 u + \tilde{G}_0^T\tilde{M}_0\zeta_2 && \text{(14c)} \\ 0 &= M^T\lambda + \tilde{M}_0^T\tilde{A}_0 z_1 + \tilde{M}_0^T\tilde{G}_0\zeta_{12} + \tilde{M}_0^T\tilde{B}_0 u + (I + \tilde{M}_0^T\tilde{M}_0)\zeta_2. && \text{(14d)} \end{aligned}$$

The matrix

$$\begin{bmatrix} I + \tilde{G}_0^T \tilde{G}_0 & \tilde{G}_0^T \tilde{M}_0 \\ \tilde{M}_0^T \tilde{G}_0 & I + \tilde{M}_0^T \tilde{M}_0 \end{bmatrix}$$

is positive definite independent of the entries of L. Hence, it is invertible and using (14c)-(14d) the noise variables can be solved for in terms of λ, z_1, and u, Substituting this expression into (14a), (14b) and redefining the coefficient matrices, the necessary conditions become (16b)–(16c).

The boundary conditions are

$$\begin{aligned} \left(z_1^T(0)P + \lambda^T(0)\right) dz_1(0) &= 0 \\ -\lambda^T(T) dz_1(T) &= 0. \end{aligned}$$

Since the initial and final states are free, the boundary conditions reduce to (16e)–(16f). Using (14c)-(14d) the integrand in (12) can now be written in terms of z_1 and u. Denote this integrand $I(z_1, u)$.

The necessary conditions for the inner minimum are taken as constraints when solving the original problem which has now been formulated as Problem 2.

From [15], a reformulation of the problem, making it easier to solve numerically, is

Problem 2.

$$\textit{Find } J = \inf_u \|u\|^2 \tag{16a}$$

$$z_1' = A_{z_1} z_1 + A_u u + A_\lambda \lambda \tag{16b}$$

$$-\lambda' = B_{z_1} z_1 + B_\lambda \lambda + B_u u \tag{16c}$$

$$\psi' = \frac{1}{2} I(z_1, u) \tag{16d}$$

with the boundary conditions

$$P_0 z_1(0) + \lambda(0) = 0 \tag{16e}$$

$$\lambda(T) = 0 \tag{16f}$$

$$\psi(0) = \frac{1}{2} z_1^T(0) P_0 z_1(0) \tag{16g}$$

$$\psi(T) \geq \gamma^2. \tag{16h}$$

In the computational example that follows we will be doing linearization around a set point. In this case z_i is the difference between the set point and the state variable.

The matrices in the dynamics are numerically computed offline prior to the minimal proper u calculation. The coordinate changes necessary to change Problem 1 into this problem do not affect the auxiliary signal and any approximations made were conservative in nature. Therefore the auxiliary signal found by solving Problem 2 will be proper for Problem 1.

Unlike Problem 1, Problem 2 can be implemented in high quality direct transcription optimal control codes. We used the Matlab based GPOPS-II [13]. In this paper, the simulations are generated from the Matlab based GPOPS-II. There are two options when solving the nonlinear programming problem that results from the discretizations. GPOPS-II comes with the interior point program IPOPT but a separate license for the sequential quadratic SNOPT is available for a fee. The results in this paper relied on IPOPT.

3. Case Study

In this paper, we examine how the work in [15] can be useful even for nonlinear models whose linearizations do not satisfy the assumptions of [15]. We consider a robot arm with elastic joints which is illustrated in Figure 1. Robot systems often use harmonic drives, belts, or long shafts as transmission elements between the motors and the links (arms) and typically display oscillations both in fast motion and after a sudden stop. Experimental tests and simulations have shown that the elasticity introduced at the joints by these transmission elements is the major reason for their vibrational behavior, so we must include elasticity in the dynamic model. As a result, the internal position of the motors does not determine the position of the driven arms. The dynamics of this displacement can be modeled by inserting a linear torsional spring at each elastic joint, between the actuator and the link. The modeling process is described in full in [7].

The model for the robot arm shown in Figure 1 is given by (17). m_0, m_1, and m_p are masses, with m_p denoting the load or object being held, l_1 and l_2 are the lengths of the arms, K is the coefficient of elasticity of joint 2, NT is the transmission ratio at the second joint, JR_i are rotor inertias, q_i are angular coordinates describing the robot's configuration, and τ_i are the rotational torques caused by the drive motors. $x = [q_1, q_2, q_3, q_1', q_2', q_3']^T$, $u_1 = \tau_1$, and

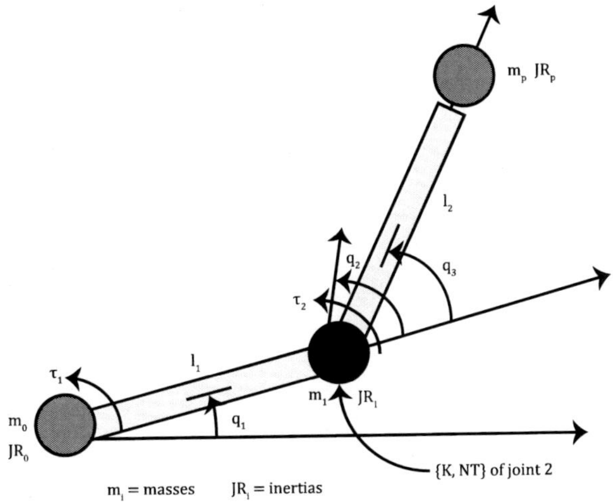

Figure 1. A two-link robot arm with the second joint elastic, from [6].

$u_2 = \tau_2$.

$$x_1' = x_4 \tag{17a}$$
$$x_2' = x_5 \tag{17b}$$
$$x_3' = x_6 \tag{17c}$$
$$x_4' = f_4(x_2, x_3, x_4, x_6) + g_{41}(x_3)u_1 - g_{41}(x_3)u_2 + \mu_1 \tag{17d}$$
$$x_5' = f_5(x_2, x_3, x_4, x_6) - g_{41}(x_3)u_1 + g_{52}(x_3)u_2 + \mu_2 \tag{17e}$$
$$x_6' = f_6(x_2, x_3, x_4, x_6) + g_{61}(x_3)u_1 - g_{61}(x_3)u_2 + \mu_3 \tag{17f}$$
$$0 = \cos x_1 + \cos(x_1 + x_3) - 1 \tag{17g}$$

During operation it is desired to have the motion of the endpoint mass be only vertical. Equation (17g) is a path constraint describing vertical motion on the endpoint of the robotic arm. This is a path constraint as opposed to a physical constraint. That is, it restricts the controls but does not generate its own forces.

The nonlinear functions g_{ij} are given by

$$\begin{aligned} g_{41}(x_3) &= \frac{A_2}{A_3(A_4 - A_3\cos^2 x_3)} \\ g_{52}(x_3) &= g_{41}(x_3) + \frac{1}{JR_1} \\ g_{61}(x_3) &= -g_{41}(x_3) - \frac{\cos x_3}{A_4 - A_3\cos^2 x_3}, \end{aligned}$$

The $f_i(x_2, x_3, x_4, x_6)$ are as follows.

$$\begin{aligned} f_4 &= \frac{A_2 \sin x_3 (x_4 + x_6)^2 + A_3 x_4^2 \sin x_3 \cos x_3}{A_4 - A_3\cos^2 x_3} \\ &+ \frac{K\left(x_3 - \frac{x_2}{NT}\right)\left(\frac{A_2}{A_3}\left(\frac{NT-1}{NT}\right) + \cos x_3\right)}{A_4 - A_3\cos^2 x_3}, \\ f_5 &= -f_4(x_2, x_3, x_4, x_6) + \frac{K}{NT}\left(x_3 - \frac{x_2}{NT}\right)\left(\frac{1}{JR_1} - 2g_{41}(x_3)\right), \\ f_6 &= -f_4 - \frac{K\left(x_3 - \frac{x_2}{NT}\right)\left(\frac{A_5}{A_3} - \left(\frac{3NT+1}{NT}\right)\cos x_3\right)}{A_4 - A_3\cos^2 x_3} \\ &- \frac{A_5 x_4^2 \sin x_3 + A_3 \sin x_3 \cos x_3 (x_4 + x_6)^2}{A_4 - A_3\cos^2 x_3}, \end{aligned}$$

and the constants are

$$\begin{aligned} A_2 &= JR_p + m_p l_2^2 \\ A_3 &= m_p l_1 l_2 \\ A_4 &= (m_1 + m_p) l_1 l_2 \\ A_5 &= (m_1 + m_p) l_1^2. \end{aligned}$$

This DAE model is nonlinear and cannot be directly used in our algorithm. It is also index three.

Fault detection in nonlinear ODE models was studied in [17]. In the ODE case, the author shows auxiliary signals can sometimes be found for the nonlinear problem by linearizing the nonlinear models and finding a minimal proper signal for the linearized models. The work in [17] shows if u is strictly proper for the linearized system and β is small enough and a positive scalar, then βu is proper for the nonlinear system with the noise bounds scaled by β^2. Furthermore, if the nonlinearities satisfy certain technical assumptions, then u is proper for the nonlinear problem, except for possibly tighter noise bounds.

Although specific to ODE models, [17] demonstrates potential for an extension to nonlinear DAE models, although there are some subtleties. In general, one would expect a linearized DAE to provide local information on the nonlinear DAE it originates from. However, linearized DAEs can feature quite different phenomena. A linearized DAE may show lower or higher index than the original DAE, suggest incorrect stability, observability, etc., and fail to remain regular [9]. On the other hand, if one linearizes around a constant solution and if some technical assumptions hold, then the vector field determined by the linear time invariant linearization approximately agrees with the vector field of actual solutions in a neighborhood around the equilibrium point [3]. The technical assumptions consist of some constant rank assumptions during a "reduction process" involving repeated differentiations and the assumed invertibility of $F_x(0, \bar{x})$ where $F(x', x) = 0$ is the nonlinear DAE and $\bar{x}$ is the equilibrium point. These assumptions frequently hold in mechanics applications such as the robot arm under consideration [10, 14].

3.1. Linearization

After a linearization, it can be shown that the model in (17) cannot meet the assumptions of [15]. One contribution of the present paper is showing how to overcome this difficulty by reducing the index of the DAE. Reducing the index of a DAE always involves some differentiation [1]. There are several ways to do this. One as shown in [2]. We do a different one here.

We differentiate the constraint (17g) twice to get two more constraints as shown in (18):

$$x_1' = x_4 \tag{18a}$$

$$x_2' = x_5 \tag{18b}$$

$$x_3' = x_6 \tag{18c}$$

$$x_4' = f_4 + g_{41}u_1 - g_{41}u_2 \tag{18d}$$

$$x_5' = f_5 - g_{41}u_1 + g_{52}u_2 \tag{18e}$$

$$x_6' = f_6 + g_{61}u_1 - g_{61}u_2 \tag{18f}$$

$$0 = \cos x_1 + \cos(x_1 + x_3) - 1 \tag{18g}$$

$$0 = -\sin(x_1)x_4 - \sin(x_1 + x_3)(x_4 + x_6) \tag{18h}$$

$$\begin{aligned} 0 = {} & -\sin x_1(f_4 + g_{41}u_1 - g_{41}u_2) - x_4^2 \cos x_1 + \\ & \sin(x_1 + x_3)\left(\frac{K(x_3 - \frac{x_2}{NT})(\frac{A_5}{A_3} - \frac{3NT+1}{NT}\cos x_3)}{A_4 - A_3\cos^2 x_3}\right. \\ & + \frac{A_5 x_4^2 \sin x_3 + A_3 \sin x_3 \cos x_3 (x_4 + x_6)^2}{A_4 - A_3\cos^2 x_3} + \\ & \left.\frac{\cos x_3}{A_4 - A_3\cos^2 x_3}u_1 - \frac{\cos x_3}{A_4 - A_3\cos^2 x_3}u_2\right) - \\ & (x_4 + x_6)^2 \cos(x_1 + x_3). \end{aligned} \tag{18i}$$

Equation (18h) is the first derivative of (18g) with respect to t and (18i) is the second derivative of (18g) with respect to t. In [2] the goal was to get index one DAE models and we used stabilized differentiation to control the extra dynamics. Here the goal is to get a system whose linearization meets the assumptions of [15]. Thus we do a very different model modification which results in a different system then that of [2].

Since the equations are solved numerically we need to reduce the constraints on the numerical solution. We eliminate (18f) and (18h). Thus the algebraic variables are now x_6, u_1. The reformulated, but equivalent, model is given in (19). This nonlinear model will now yield a linearization where the assumptions of [15] are met. This nonlinear DAE model has index 2. We drop the x_i dependencies of the functions $f_4(x_2, x_3, x_4, x_6)$, $f_5(x_2, x_3, x_4, x_6)$, and $g_{41}(x_3)$ to improve readability.

$$x_1' = x_4 \tag{19a}$$
$$x_2' = x_5 \tag{19b}$$
$$x_3' = x_6 \tag{19c}$$
$$x_4' = f_4 + g_{41}u_1 - g_{41}u_2 \tag{19d}$$
$$x_5' = f_5 - g_{41}u_1 + g_{52}u_2 \tag{19e}$$
$$0 = \cos x_1 + \cos(x_1 + x_3) - 1 \tag{19f}$$
(19g)

$$\begin{aligned}0 = &- \sin x_1(f_4 + g_{41}u_1 - g_{41}u_2) - x_4^2 \cos x_1 + \\ &\sin(x_1 + x_3)\left(\frac{K(x_3 - \frac{x_2}{NT})(\frac{A_5}{A_3} - \frac{3NT+1}{NT}\cos x_3)}{A_4 - A_3\cos^2 x_3}\right. \\ &+ \frac{A_5 x_4^2 \sin x_3 + A_3 \sin x_3 \cos x_3 (x_4 + x_6)^2}{A_4 - A_3\cos^2 x_3} + \\ &\left.\frac{\cos x_3}{A_4 - A_3\cos^2 x_3}u_1 - \frac{\cos x_3}{A_4 - A_3\cos^2 x_3}u_2\right) - \\ &(x_4 + x_6)^2 \cos(x_1 + x_3).\end{aligned} \tag{19h}$$

We observe that (19) can be rewritten

$$\bar{x}' = F(\bar{x}, w, u)$$
$$0 = G(\bar{x}, w, u)$$

where $\bar{x} = \begin{bmatrix} x_1 & x_2 & x_3 & x_4 & x_5 \end{bmatrix}^T$ are the dynamic variables, $w = \begin{bmatrix} x_6 & u_1 \end{bmatrix}^T$ are the algebraic variables, and $u = u_2$ is the auxiliary signal. The formulae for F and G are clear from (19).

The set points used in the linearized models correspond to the arm holding a mass motionless. Calculation of a set point will determine part, but not all, of u. Let u_1

be the free component. Thus, it is an algebraic variable in terms of the DAE, while u_2 is the test signal. The set points for the nominal and faulty models are found via a nonlinear solve in Matlab. The nominal model set point is found using a relatively large value for m_p to simulate an object being held. We will analyze two faults and refer to these faulty models as model 1 and model 2. Since our approach requires only two models at a time, one nominal and the other faulty, we construct two auxiliary signals, one for each faulty model paired with the nominal model. The first fault, model 1, simulates unintentionally dropping the object or realizing that the robotic arm is holding the wrong object by decreasing m_p in the faulty model. The second fault, model 2, simulates a breakdown of the second joint by lowering the elasticity parameter K. Model 0 is the nominal model. The set points are given in Table 1 and values of all parameters are given in Table 2.

Table 1. Numerically computed set points for all models

	Model 0 ($m_p = 10, K = 4$)	**Model 1** ($m_p = 1$)	**Model 2** ($K = 1$)
x_1	.57	.47	.64
x_2	1.70	1.97	1.46
x_3	.85	0.99	0.73
x_4	0	0	0
x_5	0	0	0
x_6	0	0	0
u_1	0	0	0
u_2	0	0	0

The linearization for model i is

$$\begin{aligned} E_i z_i' + F_i z_i &= B_i u + M_i \mu_i \\ y_i &= C_i z_i + N_i \eta_i \end{aligned}$$

where $z_i = \begin{bmatrix} x_1 & x_2 & x_3 & x_4 & x_5 & x_6 & u_1 \end{bmatrix}^T - \tilde{x}_i$. $u = u_2$ is the control. $\tilde{x}_i$ is a numerically computed set point from Table 1. μ_i and η_i are additive noises. The linearized models for $i = 0, 1, 2$, given below, may be calculated in Maple or Matlab.

$$E_i = \begin{bmatrix} I_{5\times 5} & 0 \\ 0 & 0 \end{bmatrix}$$

Table 2. Nominal parameter values for the robot arm

Parameter	Description	Nominal Value
K	Coefficient of elasticity at joint 2	4
NT	Transmission ratio of joint 2	2
JR_0	Rotor inertia of joint 1	1
JR_1	Rotor inertia of joint 2	1
JR_p	Inertia at endpoint	1
l_1	Length of link 1	1
l_2	Length of link 2	1
m_1	Mass of joint 2	1
m_p	Mass at endpoint	10

$$F_0 = \begin{bmatrix} 0 & 0 & 0 & -1 & 0 & 0 & 0 \\ 0 & 0 & 0 & 0 & -1 & 0 & 0 \\ 0 & 0 & 0 & 0 & 0 & -1 & 0 \\ 0 & 0.3654 & -0.7309 & 0 & 0 & 0 & -0.1660 \\ 0 & 0.3026 & -0.6053 & 0 & 0 & 0 & 0.1660 \\ 1.5239 & 0 & 0.9878 & 0 & 0 & 0 & 0 \\ 0 & -0.5578 & 1.1156 & 0 & 0 & 0 & -0.0096 \end{bmatrix}$$

$$F_1 = \begin{bmatrix} 0 & 0 & 0 & -1 & 0 & 0 & 0 \\ 0 & 0 & 0 & 0 & -1 & 0 & 0 \\ 0 & 0 & 0 & 0 & 0 & -1 & 0 \\ 0 & 1.8304 & -3.6609 & 0 & 0 & 0 & -1.1796 \\ 0 & -3.1896 & 6.3791 & 0 & 0 & 0 & 1.1796 \\ 1.4504 & 0 & 0.9939 & 0 & 0 & 0 & 0 \\ 0 & -0.7549 & 1.5098 & 0 & 0 & 0 & 0.2150 \end{bmatrix}$$

$$F_2 = \begin{bmatrix} 0 & 0 & 0 & -1 & 0 & 0 & 0 \\ 0 & 0 & 0 & 0 & -1 & 0 & 0 \\ 0 & 0 & 0 & 0 & 0 & -1 & 0 \\ 0 & 0.1189 & -0.2379 & 0 & 0 & 0 & -0.2020 \\ 0 & 0.0300 & -0.0601 & 0 & 0 & 0 & 0.2020 \\ 1.578 & 0 & 0.9801 & 0 & 0 & 0 & 0 \\ 0 & -0.2070 & 0.4138 & 0 & 0 & 0 & -0.0133 \end{bmatrix}$$

$$M_i = I, N_i = I,$$

$$B_0 = \begin{bmatrix} 0 \\ 0 \\ 0 \\ -0.1660 \\ 1.1660 \\ 0 \\ -0.0096 \end{bmatrix}, \quad B_1 = \begin{bmatrix} 0 \\ 0 \\ 0 \\ -1.1796 \\ 2.1796 \\ 0 \\ 0.2150 \end{bmatrix}, \quad B_2 = \begin{bmatrix} 0 \\ 0 \\ 0 \\ -0.202 \\ 1.2020 \\ 0 \\ -0.0133 \end{bmatrix}$$

$$C_0 = \begin{bmatrix} 1 & 0 & 0 & 0 & 0 & 1 & 0 \\ 0 & 0 & 1 & 0 & 0 & 0 & 0 \end{bmatrix},$$

$$C_1 = C_2 = \begin{bmatrix} 1 & 0 & 0 & 0 & 0 & 0 & 0 \\ 0 & 0 & 1 & 0 & 0 & 1 & 0 \end{bmatrix}.$$

Note that the faults are both a change in m_p or K and a change in the structure of the output matrix C.

The optimal auxiliary signals found by solving Problem 2 with a noise bound of $\gamma = 0.25$ are given in Figure 2. Their similarity is a symptom of the effects the faults have on the system. Changes in mass or elasticity will both result in an unexpected vertical displacement of the endpoint. Applying a burst of torque, which is essentially what the system is forced to do given a short interval like $[0 \;\; 1]$, will expose the problem in either case. However, it is interesting to observe the minimal proper signals over a longer interval, $[0 \;\; 10]$, shown in Figure 3. Here, we see a definite difference in the signals because on a longer interval the software can consider smaller signals that exploit more subtle differences such as resonant frequencies.

It is important to note that our original nonlinear model did not have uncertainty in it. The uncertainty was added to the linear model after linearization. Since differentiations are involved in the reformulation it can be more difficult to always introduce the uncertainty on the original nonlinear model. Thus it is important to evaluate how well the signal from the linear model works on the original nonlinear problem with uncertainty. The results of the next section show that at least for this example a useful test signal is found by adding the uncertainty to the linear problem.

4. Evaluation of the Test Signal

Given a test signal u for the linearized problem, we want to examine its use on the nonlinear problem. The ideas of this section are useful in evaluating a test signal found by any method such as that of [2]. In [17], experimentation showed that a signal found for the linear problem with a noise bound of one

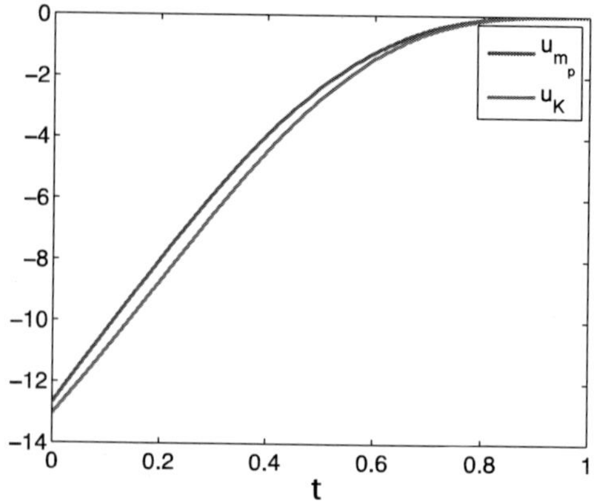

Figure 2. Comparison of numerically computed minimal proper signals on $[0\ \ 1]$. u_{m_p} is the signal for models 0 and 1, and u_K is the signal for models 0 and 2.

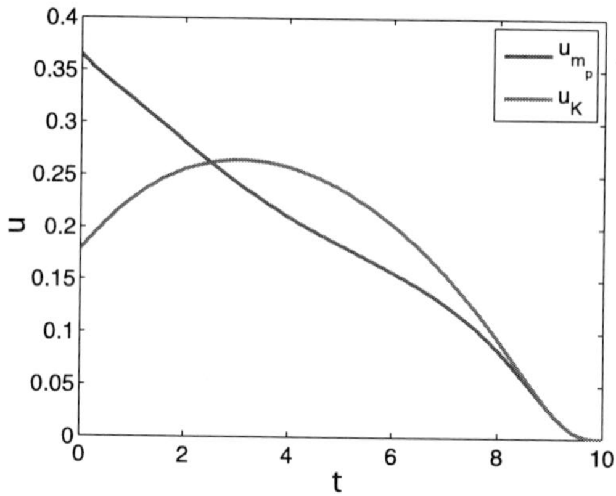

Figure 3. Comparison of numerically computed, minimal proper signals u_{m_p} and u_K on $[0\ \ 10]$.

is usually proper for the nonlinear problem, except the noise bound may have to be scaled from one. To test the signal, we fix u to be the minimal, proper signal for the linear problem and found the smallest noise necessary to satisfy the nonlinear constraints and equal outputs. The scaled noise bounds for the nonlinear test problems ranged from 0.708 to 1.117. Note, if the noise bound is found to be greater than one, then u found for the linear problem is still proper for the nonlinear problem with noise bound one, just not minimal proper.

The analogous problem here is

$$\min_{\eta_0,\eta_1,\mu_0,\mu_1,w_0,w_1} \delta \tag{20a}$$

$$\bar{x}_0' = F_0(\bar{x}_0, w_0, u) + \bar{\mu}_0 \tag{20b}$$
$$\bar{x}_1' = F_1(\bar{x}_1, w_1, u) + \bar{\mu}_1 \tag{20c}$$

$$\omega' = \eta_0^T\eta_0 + \bar{\mu}_0^T\bar{\mu}_0 + \tilde{\mu}_0^T\tilde{\mu}_0 + \eta_1^T\eta_1 + \bar{\mu}_1^T\bar{\mu}_1 + \tilde{\mu}_1^T\tilde{\mu}_1, \tag{20d}$$
$$0 = \omega(0) \tag{20e}$$
$$0 = G_0(\bar{x}_0, w_0, u) + \tilde{\mu}_0 \tag{20f}$$
$$0 = G_1(\bar{x}_1, w_1, u) + \tilde{\mu}_1 \tag{20g}$$
$$0 = C_0[\bar{x}_0 \;\; w_0] + \eta_0 - (C_1[\bar{x}_1 \;\; w_1] + \eta_1) \tag{20h}$$
$$\delta \geq \frac{1}{2}(\bar{x}_0^T(0) - \tilde{x}_0)\hat{P}_0(\bar{x}_0(0) - \tilde{x}_0) + \frac{1}{2}(\bar{x}_1^T(0) - \tilde{x}_1)\hat{P}_1(\bar{x}_1(0) - \tilde{x}_1) + \omega(T), \tag{20i}$$

Here $\tilde{x}_i$ is the set point for model i and $\mu_i = [\bar{\mu}_i \;\; \tilde{\mu}_i]$. Barred variables are original variables. The subscript i denotes which model, $i = 0$ (nominal) or $i = 1, 2$ (faulty), is being considered. We have also simplified the equations by ignoring the coefficient matrices on the noises since they are identity matrices in our example. u is either u_{m_p} or u_K from Figure 2. We can solve for some of the noise using equations (20f)–(20h) and rewrite the problem

$$\min_{\eta_1,\bar{\mu}_0,\bar{\mu}_1,w_0,w_1} \delta \tag{21a}$$
$$\bar{x}_0' = F_0(\bar{x}_0, w_0, u) + \bar{\mu}_0 \tag{21b}$$
$$\bar{x}_1' = F_1(\bar{x}_1, w_1, u) + \bar{\mu}_1 \tag{21c}$$
$$\omega' = (C_1[\bar{x}_1 \;\; w_1] + \eta_1 - C_0[\bar{x}_0 \;\; w_0])^T (C_1[\bar{x}_1 \;\; w_1] + \eta_1 - C_0[\bar{x}_0 \;\; w_0]) + \bar{\mu}_0^T\bar{\mu}_0 + G_0^TG_0 + \eta_1^T\eta_1 + \bar{\mu}_1^T\bar{\mu}_1 + G_1^TG_1, \;\; \omega(0) = 0 \tag{21d}$$
$$\delta \geq \frac{1}{2}(\bar{x}_0^T(0) - \tilde{x}_0^T)\hat{P}_0(\bar{x}_0(0) - \tilde{x}_0) + \frac{1}{2}(\bar{x}_1^T(0) - \tilde{x}_1^T)\hat{P}_1(\bar{x}_1(0) - \tilde{x}_1) + \omega(T). \tag{21e}$$

Note that if Model 2 is the faulty model, then the 1 subscript in (20) and (21) becomes a 2.

Let $\bar{\Gamma}_1$ denote the calculated minimum value of Γ for the pair of models 0 and 1 and $\bar{\Gamma}_2$ denote the minimum value of Γ for models 0 and 2. Our simulations estimated $\Gamma_1^2 = .0672$ and $\Gamma_2^2 = .0360$. Recall, $\gamma = 0.25$ and note $\sqrt{\Gamma_1^2} = 0.259$, indicating u_{m_p} is proper on the nonlinear problem for essentially the same noise bound as for the linearized problem. $\sqrt{\Gamma_2^2} = 0.19$, thus the linear test signal u_K found with a noise bound of $\gamma = 0.25$, would work well on the nonlinear problem, but the signal would provide perfect detection only with the smaller additive noise bound of 0.19. This example shows the utility of our algorithm on practical problems with models and faults that come from real-world applications, even ones with nonlinear phenomena. Finding a proper u was successful by utilizing linearized models.

5. Model Identification with Nonlinear Models

To identify which model is active, the natural problem to solve is

$$W_i(T) = \min_{\eta_i, \bar{\mu}_i, \tilde{\mu}_i} \frac{1}{2}\left(\bar{x}_i^T(0) - \tilde{x}_i^T\right)\hat{P}_i\left(\bar{x}_i(0) - \tilde{x}_i\right) + \frac{1}{2}\int_0^T \|\eta_i\|^2 + \|\mu_i\|^2\, dt \tag{22a}$$

$$\bar{x}_i' = F_i(\bar{x}_i, w_i, u) + \bar{\mu}_i \tag{22b}$$

$$0 = G_i(\bar{x}_i, w_i, u) + \tilde{\mu}_i \tag{22c}$$

$$y = C_i\left[\bar{x}_i \;\; w_i\right] + \eta_i, \tag{22d}$$

where y typically comes from sensor readings. Note that $W_i(t)$ is monotonically nondecreasing and if $W_i(t) > \gamma^2$ for some t, then model i is not realizable. Solving this problem for $i = 0, 1$ for $t = T$ will successfully identify the active model and is easily formulated but will be difficult to do in real time for high dimensional complex systems if that is required.

For linear models there is a fast way to compute the W_i as the y_i are measured. We now look at whether we can use the linearizations and the linear model identification approach from [15] on problems with nonlinear phenomena. As with any linearized model, its accuracy and utility may suffer as the states and controls move away from the operating point. Auxiliary signals, by their very definition, briefly force systems off of their operating points to expose faults. This means the noise estimation technique in Theorem 3.1 of [15] will estimate the model error of the linearized model as well as the noise required to generate the observed output. The result could be imperfect detection.

As noted there are several positive results on when linearizations can be used on nonlinear DAEs [3, 10, 14]. These results cover a variety of constrained mechanical systems such as the example in this paper. For example, if the technical assumptions of [3] hold, then the vector field determined by the linear time invariant linearization approximately agrees with the vector field of actual solutions in a neighborhood around the equilibrium points. Applications with small noise bounds and small proper signals may also benefit from this approach. Small noises and signals will keep the states from straying too far from the operating point, cutting the noise generated from modeling error. One can also modify the cost of the test signal to include the magnitude of the state displacement during the test [4].

In practice, y comes from sensor data. In this paper we numerically simulate a nonlinear model to obtain y. Note that some controls will cause the arm to move through kinematic singularities, for example, when $x_1 = x_3 = 0$. This is merely a simulation problem and not a problem of the approach because in practice there is no need to simulate the nonlinear model.

Smaller noises and smaller signals will mitigate this problem. They are also more reasonable in our example. The noise bound in the previous section, $\gamma = 1/4$, is a significant amount of noise in terms of the angles, which are in radians. The following figures result from the computation of a new proper signal for models 0 and 1, and 0 and 2 (see Table 1) with $\gamma = \frac{1}{70}$. The test signal is given in Figure 4. We generate noise that meets the bound and simulate the nominal model ($i = 0$) to find y. As y is computed we calculate the past noise required to generate y for each linear model.

The results are presented in Figures 5 and 6. The noise estimation correctly identifies model 0 as being the correct model in both cases before $t = 5$. Here, the system stays close to its operating point and the linearizations are good approximations of the nonlinear systems. The result is good detection. Even by experimenting with higher noise bounds and larger signals, we were unable to find an instance of failed detection, so long as the system was able to be simulated.

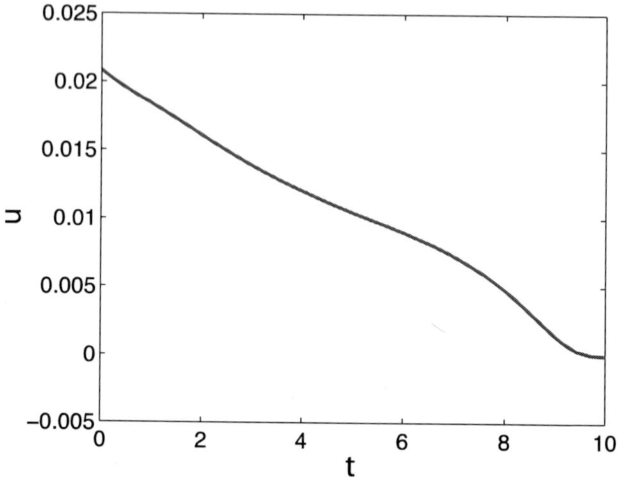

Figure 4. Proper auxiliary signal separating models 0 and 1 with $\gamma = \frac{1}{70}$.

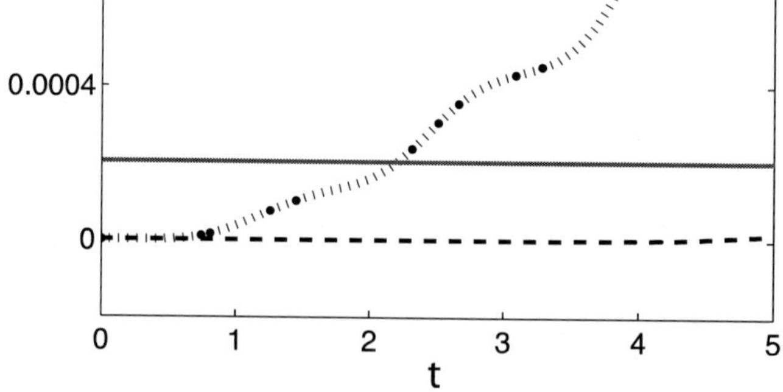

Figure 5. $W_i(t)$ and γ^2 for fault 1 with Model 0 simulated.

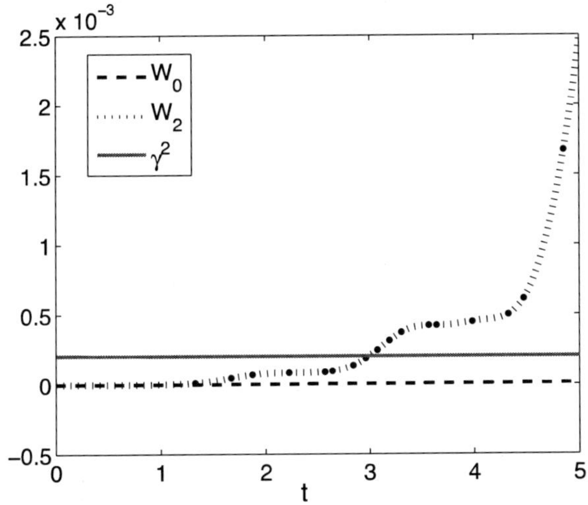

Figure 6. $W_i(t)$ and γ^2 for fault 2 with Model 0 simulated.

Conclusion

In this paper we have examined the application of a previously developed fault detection technique for linear DAE models, to the linearization of a nonlinear DAE robot arm model. Useful signals were computed from linearizations of the robot arm model using this prior theory. Moreover, fault identification theory, that is, deciding which model is present after administering the auxiliary signal, was also successful.

For the approach of [15] to be used there are some technical assumptions that must be met. Steps to modify a problem that does not meet these assumptions, to one that does meet the assumptions, is carefully presented in the context of a highly nonlinear application. The assumptions were not initially met by the linearization of the robot arm but they were after a model modification and reduction to index two. It is important to note that we did not need to reduce the index to one.

An algorithm for determining the validity of the test signal on the full nonlinear problem was presented. The algorithm is in the form of a constrained nonlinear optimization problem that can be solved by existing available software. In the case of our example application it showed that we did not have to introduce the uncertainty on the original nonlinear problem but could wait and introduce it on the linearization provided the amount of uncertainty was not too large.

The various modeling and computational issues in the application of this approach were carefully discussed.

Acknowledgment

Research supported in part by NSF Grant DMS-1209251.

References

[1] K. Brenan, S. L. Campbell, and L. R. Petzold. *Numerical Solution of Initial Value Problems in Differential-Algebraic Equations*. SIAM, Philadelphia, PA, 1996.

[2] S. Campbell. Active fault detection in nonlinear differential algebraic equations i: General systems. Preprint, 2016.

[3] S. L. Campbell. Linearization of DAEs along trajectories. *Z. angew Mathematical Physics (ZAMP)*, 46:70–84, 1995.

[4] S. L. Campbell and R. Nikoukhah. *Auxiliary Signal Design for Failure Detection*. Princeton University Press, Princeton, New Jersey, 2004.

[5] D. Choe, S. L. Campbell, and R. Nikoukhah. A comparison of optimal and suboptimal auxiliary signal design approaches for robust failure detection. In *Proc. IEEE Conference on Control Applications*, Toronto, 2005.

[6] De Luca, A. Control properties of robot arms with joint elasticity. In *Proc. International Symposium on the Mathematical Theory of Networks and Systems*, Phoenix, AZ, June 1987.

[7] De Luca, A. and Isidori, A. Feedback linearization of invertible systems. In *2nd Duisburger Kolloquium Automation und Robotik*, Duisburg, Germany, July 1987.

[8] P. Kunkel and V. Mehrmann. *Differential-Algebraic Equations. Analysis and Numerical Solution*. EMS Publishing House, Zürich, Switzerland, 2006.

[9] R. März. Notes on linearization of DAEs and on optimization with differential-algebraic constraints. In L. T. Biegler, S. L. Campbell, and V. Mehrmann, editors, *Control and Optimiation with Differential-Algebraic Constraints*, Advances in Design and Control, chapter 3, pages 37–58. SIAM, 2012.

[10] N. H. McClamroch. Feedback stabilization of control systems described by a class of nonlinear differential-algebraic equations. *Systems and Control Letters*, 15:53–60, 1990.

[11] H. H. Niemann. A setup for active fault diagnosis. *IEEE Transactions on Automatic Control*, 51(9):1572–1578, 2006.

[12] N. Poulsen and H. H. Niemann. Active fault diagnosis based on stochastic tests. *International Journal of Applied Mathematics and Computer Science*, 18:487–496, 2008.

[13] A. V. Rao, D. A. Benson, C. Darby, M. A. Patterson, C. Francolin, I. Sanders, and G. T. Huntington. Algorithm 902: Gpops, a matlab software for solving multiple-phase optimal control problems using the gauss pseudospectral method. *ACM Transactions on Mathematical Software*, 37(37):22:1–22:39, 2010.

[14] S. Reich. On the local qualitative behavior of differential-algebraic equations. *Circuits, Systems, and Signal Processing*, 14:427–443, 1995.

[15] J. R. Scott and S. L. Campbell. Auxiliary signal design for failure detection in differential-algebraic equations. *Numerical Algebra, Control and Optimization*, 4:151–179, 2014.

[16] J. R. Scott and S. L. Campbell. Auxiliary signal design for failure detection in high index differential-algebraic equations. In *Proc. 2014 IEEE Conference on Decision and Control*, Los Angeles, CA, 2014.

[17] K. Sweetingham and S. Campbell. Auxiliary signal design for fault detection in nonlinear systems. *Control and Intelligent Systems*, 37:152–159, 2009.

In: Fault Detection
Editor: Daniel Martin

ISBN: 978-1-53610-345-8
© 2017 Nova Science Publishers, Inc.

Chapter 3

SEISMIC ATTRIBUTE-AIDED FAULT DETECTION IN PETROLEUM INDUSTRY: A REVIEW

Haibin Di* and Dengliang Gao†
[1]Formerly Department of Geology and Geography,
West Virginia University, Morgantown, West Virginia, US
presently Center for Energy and Geo Processing (CeGP),
Georgia Institute of Technology, Atlanta, Georgia, US

Abstract

In petroleum exploration and production, faults are of importance by controlling pathways of hydrocarbon migration and accumulation in the subsurface, and robust fault detection is one of the major tasks of geologic and geophysical interpretation from three-dimensional (3D) seismic data. Traditionally, faults are interpreted by manually picking on vertical/horizontal seismic sections with geologic consistence; however, such manual tool is time consuming and sensitive to the visibility and interpreters' bias, especially for a large dataset with structural complexities. With the development of new signal processing and data visualization technologies, computer-aided semi-automatic/automatic fault extraction has been the focus of recent geophysical research on seismic fault detection with superiorities in both computational efficiency and result accuracy. Various methods have been developed and

* E-mail address: haibin.di@ece.gatech.edu.
† E-mail address: dengliang.gao@mail.wvu.edu.

implemented in the petroleum industry; however, their results are often limited by the detection resolution and contaminated by non-fault artifacts that interfere with reliable structural interpretation. This paper provides a comparative analysis and comprehensive review of the existing fault detection methods, including manual picking and computer-aided automatic/semi-automatic extraction. A suite of algorithms for each method are tested regarding their computational efficiency and result accuracy by calibrating their results to the original seismic images for quality control. Finally, for addressing the limitations and problems observed in the existing techniques, innovative solutions are presented to further enhance the reliability of fault interpretation from seismic data.

Introduction

Fault detection from three-dimensional (3D) seismic data is essential for subsurface structure interpretation and reservoir characterization in petroleum exploration and production. Traditionally, faults are manually picked from vertical sections of seismic amplitude by an experienced interpreter. However, manual picking is time consuming, labor intensive, and moreover sensitive to interpreters' bias, especially for large datasets with complex fracture networks. Then, in order to improve both interpretational efficiency and accuracy, geologists and geophysicists have made great efforts in developing algorithms for effective semi-automatic/automatic fault extraction from 3D seismic data with the aid of powerful computers and workstations (e.g., Pedersen et al., 2004; Tingdahl & de Rooij, 2005; Admasu et al., 2006; Hale, 2013; Wang and AlRegib, 2014b; Zhang et al., 2014). In general, these algorithms consist of three steps: first to generate a suite of seismic attributes (such as discontinuity) from the seismic dataset, in which faults are highlighted as lineaments; then to perform fault enhancement that further sharpens the imaging of seismic faults with the non-fault features suppressed or even removed, so that they could be easily differentiated by computer programs; finally to extract patches from the enhanced fault images by either an automatic detection algorithm with fault-related parameter configurations (such as fault size and orientation) or a semi-automatic detection algorithm with seeds pre-defined by an interpreter.

3D seismic surveying has been the most popular method for reservoir exploration and production since it was first applied in 1970s owing to its straightforward delineation of subsurface geologies by monitoring seismic wave propagation in three dimensions. In particular, faults are associated with abrupt lateral changes in seismic amplitude and/or waveform, and such observation allows geophysicists to highlight the faults by evaluating such changes in a qualitative/quantitative manner, denoted as discontinuity attribute. Seismic coherence is the first discontinuity attribute presented for detecting faults and stratigraphic features that uses a cross-correlation operator to estimate lateral waveform similarity (Bahorich and Farmer, 1995), whereas lateral variation of seismic amplitude was first estimated by applying the Sobel filter (Luo et al., 1996). The first-generation algorithms involve only three neighboring traces, making them highly sensitive to noises present in seismic data. The signal/noise ratio of the generated discontinuity images is improved by incorporating more traces into discontinuity analysis, such as the semblance presented by Marfurt et al. (1998), the eigenstructure-based coherence presented by Gersztenkorn and Marfurt (1999), the local structural entropy presented by Cohen and Coifman (2002), and the similarity presented by Tingdahl and de Rooij (2005).In general, all these algorithms are based on the assumption of seismic signals being zero mean, from which faults can be highlighted in an accurate manner. In most cases, however, mean of localized amplitude/waveform variation rarely is zero; thereby performing discontinuity detection on such data will undesirably weaken the contrast between discontinuities and continuous portions and decrease the detection resolution. The limitation is addressed by rescaling local amplitude before discontinuity attribute computation, so that the lateral resolution could be enhanced for subtle faults and fractures existing in seismic reflectors with nonzero-mean amplitude variation (Di and Gao, 2014). The discontinuity analysis greatly facilitates manual fault picking, and seismic faults especially those perpendicular to the beddings can be clearly recognized and picked by interpreters. However, when applying the generated discontinuity images to semi-automatic/automatic fault extraction, they are limited in both resolution and accuracy. First, due to noises in seismic data, discontinuity attributes highlight not only the true faults of interpretation interest, but also some artifacts with discernable waveform/amplitude variations that may be undesirably extracted as fault patches by computers. Second, for faults with a wide damaging zone, discontinuity attributes often highlight them as lineaments with a certain thickness, instead of thin lines, which adds to the difficulties for computers to determine the optimal location for extracting a one-pixel-thick patch that best represent such a thick faulting zone.

Then, fault enhancement is presented for improving the discontinuity resolution and moreover thinning the thick lineaments, and various approaches have been developed and applied to seismic fault detection (e.g., Pedersen et al., 2004; Cohen et al., 2006; Bag & Harit, 2011; Hale, 2013; Wang et al., 2014; Zhang et al., 2014).Among them, the ant tracking (Pedersen et al., 2004) and the lineament thinning (Hale, 2013) are considered most popular with wide applications in the industry. In particular, the former applies the principles of swarm intelligence and describes the collective behavior of ants finding the shortest path between the nest and a food source by communicating via pheromone, a chemical substance that attracts other ants (Pedersen et al., 2004).However, while enhancing the faults in seismic data, such tracking also exaggerates artifacts with weak waveform/amplitude variations, such as processing effects, channel boundaries, and chaotic responses, all of which are magnified to the same level as the faults of interest and could be extracted as fault patches by mistake. Instead, the thinning approach proposed by Hale (2013) is capable of generating one-pixel-thick lineaments without exaggerating artificial lineaments by performing discrete scanning over both fault strike and dip to search for maximum waveform semblance. Correspondingly, the thinned lineaments are capable of accurately representing seismic faults and thus can be easily identified as patches by computers without causing ambiguities in determining fundamental fault properties (e.g., location, size, and orientation).

Manual picking of seismic objects, such as faults, horizons and salt domes, is limited by its efficiency and sensitivity to interpreter bias, especially for large seismic datasets with interpretational uncertainties; then semi-automatic/automatic extraction has been the research focus since 1990s. For example, Meldahl et al. (1999) presented a semi-automatic approach for detecting seismic chimneys by combining directive attributes and neural network, and such approach was later adapted to fault extraction from 3D seismic data by Tingdahl and de Rooij (2005). Gibson et al. (2005) and Zhang et al. (2014) proposed grouping fault points into the local planar patches and then merging these small patches into larger fault surfaces under certain geometric constraints. Admasu et al. (2006) proposed an autotracking method of fault line propagation from one section to another within the whole volume to obtain fault patches. Hale (2013) proposed a dynamic time warping algorithm to generate fault surfaces based on the boundary constraints derived from the thinned discontinuity images. Wang and AlRegib (2014a, b) introduced 3D Hough transform to detect potential fault planes in a seismic volume and the ideas of motion vectors in video coding and processing to extract fault surfaces. Unfortunately, all these algorithms are unable to simultaneously achieve both high resolution in extracting true faults and high accuracy in avoiding non-fault artifacts. The common result is either an aggressive case with high resolution (most/all true faults extracted) but low accuracy (many artifacts introduced), or a conservative one with high accuracy (few artifacts introduced) but low resolution (few true faults extracted). Therefore, semi-automatic/automatic fault extraction is still in the experimental phase for testing and not ready for practical implementation and application to industrial projects.

The purpose of this paper is to provide a comprehensive review for fault interpretation techniques from 3D seismic data presently used in the petroleum industry. The paper starts with a description of the seismic attributes that are most useful for fault delineation and interpretation; then fault detection is performed by both manual picking and semi-automatic/automatic extraction using a 3D seismic dataset from the North Sea that is rich in natural faults. Finally, the results are calibrated to the original seismic images for quality control, and future developments are proposed for overcoming the observed problems and leading to more robust computer-aided detection of seismic faults.

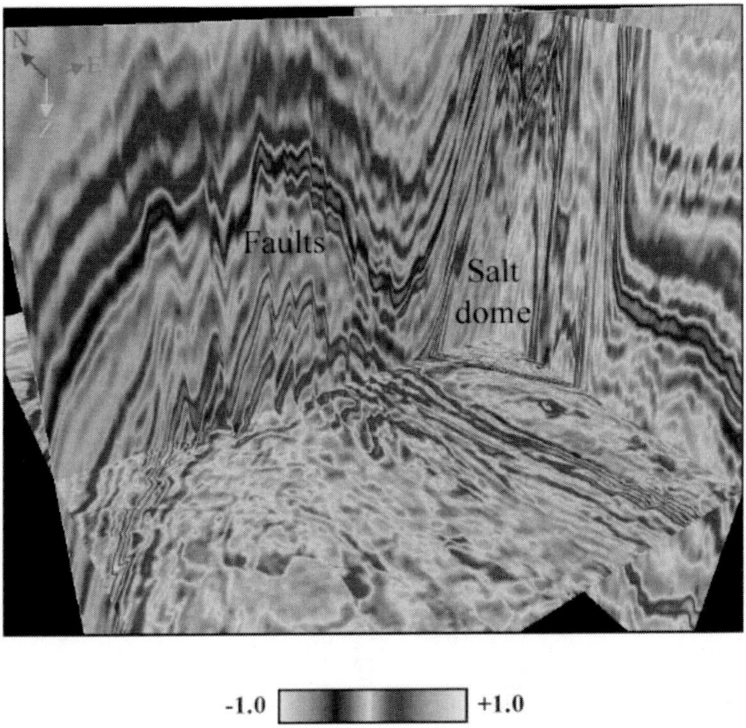

Figure 1. The 3D seismic dataset from the North Sea dominated by salt domes and salt-related faults.

Attribute Extraction

This section summarizes the major approaches implemented in the fault-detection algorithms that are capable of promoting the visibility of seismic faults, including data preconditioning, discontinuity analysis, and fault enhancement. To be consistent for comparative analysis, we use a 3D seismic volume from the North Sea (Figure 1) that is dominated by salt domes and associated faults. The dataset contains 600 inlines and 950 crosslines with the depth ranging from 1552 ms to 1848 ms at 4-ms interval, and all results are shown along the same vertical and horizontal sections for fair comparisons.

Post-stack amplitude is the most widely used seismic data in subsurface exploration. Figure 2 displays the horizontal slice of 1728 ms and the vertical slice of inline 425 in the dataset, with the black line indicating the location of the corresponding slice. From both slices, several faults are recognized as abrupt changes in the waveform patterns, especially for those perpendicular to the beddings. In contrast, the bedding-parallel faults are not well revealed in the amplitude images, and highlighting such fault usually requires the assistance with seismic attribute analysis, such as discontinuity.

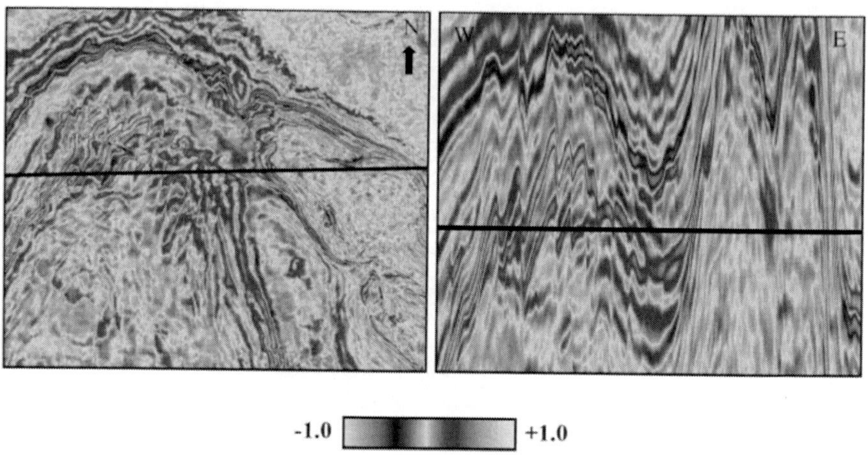

Figure 2. The horizontal slice of 1728 ms (left) and the vertical slice of inline 425 (right) of post-stack amplitude. Black line denotes the location of one in the other.

Data Preconditioning

Due to the noises existing in 3D seismic data, noise-suppression techniques, such as signal filtering and data smoothing, are often applied for data conditioning before interpretation, so that the features of interpretation interest could be shown more clearly whereas the noise-related artifacts could be effectively suppressed or even eliminated. Various filtering/smoothing algorithms have been introduced from other fields like image processing, and two most popular ones are median filter and edge-preserved smoothing.

Median Filter

Median filter is a nonlinear smoothing method that was first developed in image processing; and when applied to seismic data filtering, it works effectively for removing random noises and at the same time preserving seismic discontinuities. Such filter is easy for implementation by replacing the value at a sample with the median value of neighboring samples within a user-defined 2D or 3D analysis window. Figure 3 displays the corresponding horizontal slice of 1728 ms and the vertical slice of inline 425. Compared to the original post-stack amplitude (Figure 2), both images are cleaned after median filtering with the faults better highlighted, especially in the vertical section.

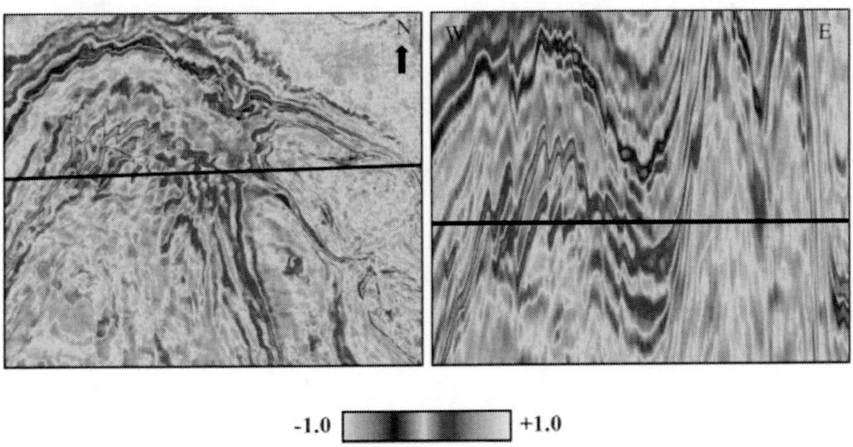

Figure 3. The horizontal slice of 1728 ms (left) and the vertical slice of inline 425 (right) of median filtering. Black line denotes the location of one in the other.

Edge-Preserving Smoothing

Edge-preserving smoothing is a derivation of the general running-average smoothing method and effectively resolves the conflict between noise suppression and edge degradation from the conventional statistical filters, such as the median filter. The principle of such method is to search for the most homogeneous fragment around each sample within a user-defined 2D or 3D window and assigns the average value of the fragment to that sample. The standard deviation is often selected for quantifying local homogeneity (Luo et al., 2002). However, such discrete searching increases the associated computational expense especially when the searching is in 3D and the window size is large. Figure 4 displays the corresponding horizontal slice of 1728ms and the vertical slice of inline 425, in which the noises are significantly suppressed and more importantly, the faults are enhanced with their planes delineated more clearly than those in Figure 3.

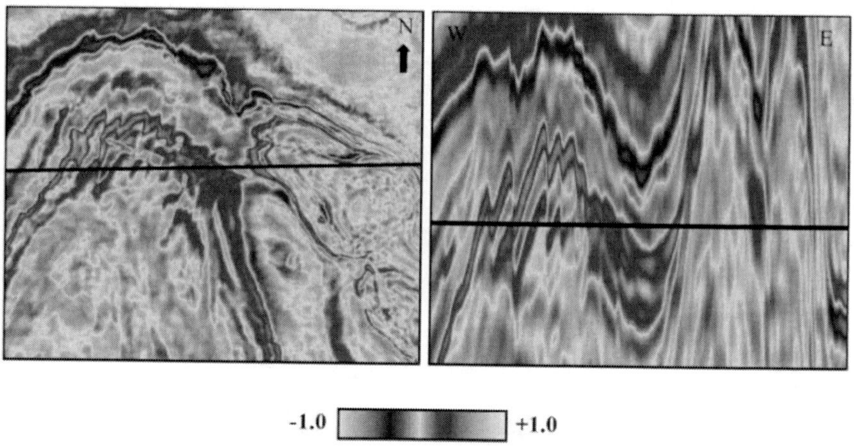

Figure 4. The horizontal slice of 1728 ms (left) and the vertical slice of inline 425 (right) of edge-preserving smoothing. Black line denotes the location of one in the other.

Discontinuity Analysis

Direct fault interpretation from seismic amplitude is limited in two major ways. The first limitation is its low resolution on highlighting faults that are parallel/subparallel to seismic reflections, especially when the dataset is contaminated by noises. Second, amplitude is applicable to only manual fault interpretation; when turning to semi-automatic/ automatic fault extraction, such information is not uniquely linked to faults although it could be well resolved by interpreters. Instead, computer-aided fault detection requires an edge-type display of seismic faults as input, in which only faults are highlighted whereas other features are strongly depressed or even removed. Both limitations are addressed by performing discontinuity analysis from seismic amplitude, which quantifies the lateral variations of seismic amplitude and/or waveform. There exists a suite of discontinuity attributes, and the major ones are selected as follows.

Coherence

Seismic coherence was the first discontinuity attribute presented for delineating faults and stratigraphic features, which uses a three-trace cross-correlation operator to measure the similarity between the localized waveform at a sample and those at its surrounding samples (Bahorich and Farmer, 1995). Let s_0 denote the waveform at the target sample, s_x and s_y the waveforms at its surrounding sample along the inline (x-) and crossline (y-) directions, respectively, then the coherence attribute c_1 is estimated as

$$c_1 = \frac{\sum_{k=1}^{K}(s_0(k)-\overline{s_0})\cdot(s_x(k)-\overline{s_x})}{\sqrt{\sum_{k=1}^{K}(s_0(k)-\overline{s_0})^2}\cdot\sqrt{\sum_{k=1}^{N}(s_x(k)-\overline{s_x})^2}}\cdot\frac{\sum_{k=1}^{K}(s_0(k)-\overline{s_0})\cdot(s_y(k)-\overline{s_y})}{\sqrt{\sum_{k=1}^{K}(s_0(k)-\overline{s_0})^2}\cdot\sqrt{\sum_{k=1}^{N}(s_y(k)-\overline{s_y})^2}} \quad (1)$$

where $\overline{s_0}$, $\overline{s_x}$ and $\overline{s_y}$ denote the average of waveforms s_0, s_x and s_y, respectively.K denotes the size of vertical analysis window.

The visualization of seismic coherence often uses a black-gray-white color map, in which a fault is highlighted in black (low value) indicating significant waveform changes, whereas the continuous portion is in white (high value). Figure 5 displays the coherence images of the horizontal slice of 1728 ms and the vertical slice of inline 425, in which the faults are clearly detected especially those parallel to beddings (denoted by arrows).

Sobel Filtering

Sobel filter is a common edge detector used in the field of image processing, which effectively detects edges in an image by measuring the gradient of amplitude at each pixel. When applied to seismic data, the amplitude gradient c_2 is estimated as the square root of two gradients of seismic amplitude u along inline (x-) and crossline (y-) directions (Luo et al., 1996),

$$c_2 = \sqrt{\left(\frac{\partial u}{\partial x}\right)^2 + \left(\frac{\partial u}{\partial y}\right)^2} \quad (2)$$

In contrast to the coherence attribute, amplitude gradient highlights faults with high values, whereas the continuous portions have low amplitude gradient. For comparison, the color polarity for amplitude gradient is reversed in this paper, so that a fault is always associated with black whereas continuous structure is white. Figure 6 displays the amplitude-gradient images of the horizontal slice of 1728ms and the vertical slice of inline 425, in which the faults are also clearly imaged in the way similar to the coherence attribute (Figure 5).

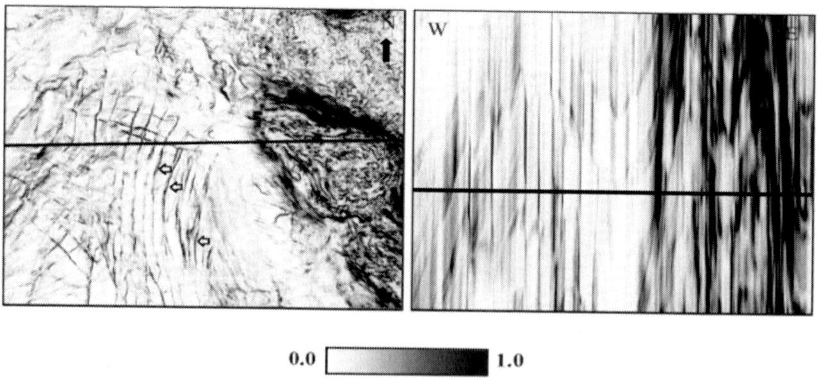

Figure 5. The horizontal slice of 1728 ms (left) and the vertical slice of inline 425 (right) of coherence.

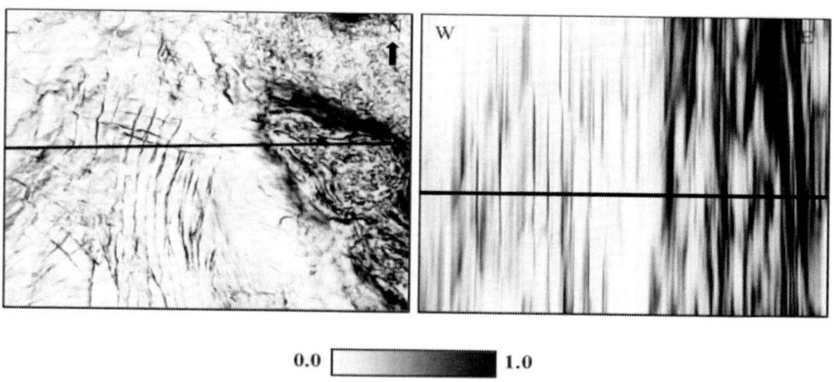

Figure 6. The horizontal slice of 1728 ms (left) and the vertical slice of inline 425 (right) of Sobel filtering.

Semblance

The semblance attribute measures lateral changes in seismic amplitude by incorporating more traces to reduce the sensitivity to noise present in a seismic dataset (Marfurt et al., 1998). Let u denote the amplitude within a lateral analysis window of J samples and a vertical analysis window of K samples, then the semblance c_3 is estimated as

$$c_3 = \frac{\sum_{k=1}^{K}\left\{\left[\sum_{j=1}^{J} u(j,k)\right]^2 + \left[\sum_{j=1}^{J} u_H(j,k)\right]^2\right\}}{J\sum_{k=1}^{K}\sum_{j=1}^{J}\left[u^2(j,k)+u_H^2(j,k)\right]} \tag{3}$$

where u_H denotes the Hilbert transform or quadrature component of the real seismic amplitude u . The quadrature amplitude allows to obtain robust estimates of semblance even about the zero crossings of seismic reflection events (Marfurt et al., 1998).

Similar to the coherence attribute, faults are highlighted in black (low semblance) as amplitude variations within the defined analysis windows are assumed being zero mean, whereas the continuous portions are in white (high semblance) due to theoretically constant amplitude in the windows. Figure 7 displays the semblance images of the horizontal slice of1728 ms and the vertical slice of inline 425, in which the faults are clearly imaged with higher resolution on the portions associated with weak waveform/amplitude variation (denoted by the black circle).

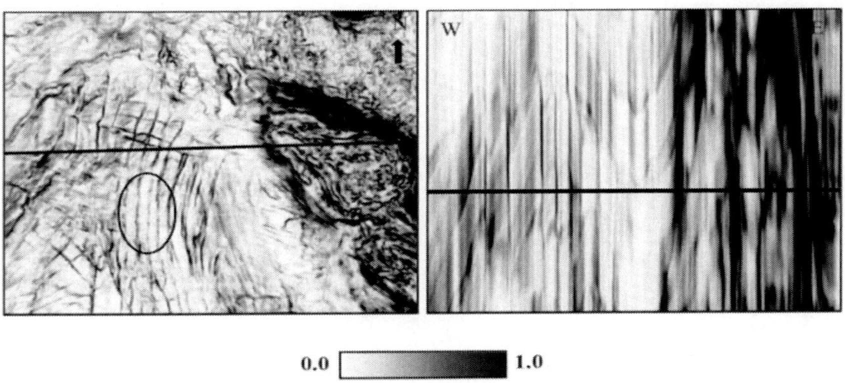

Figure 7. The horizontal slice of 1728 ms (left) and the vertical slice of inline 425 (right) of semblance.

Similarity

The similarity attribute integrates the waveform similarity and amplitude variation, and the three-trace similarity c_4 is estimated as (Tingdahl and de Rooij, 2005)

$$c_4 = \frac{1}{2}\left[\frac{\sqrt{\sum_{k=1}^{K}[s_0(k)-s_x(k)]^2}}{\sqrt{\sum_{k=1}^{K} s_0{}^2(k)}+\sqrt{\sum_{k=1}^{K} s_x{}^2(k)}} + \frac{\sqrt{\sum_{k=1}^{K}[s_0(k)-s_y(k)]^2}}{\sqrt{\sum_{k=1}^{K} s_0{}^2(k)}+\sqrt{\sum_{k=1}^{K} s_y{}^2(k)}}\right] \quad (4)$$

where s_0 denotes the waveform at the target sample. s_x and s_y the waveforms at its surrounding sample along the inline (x-) and crossline (y-) directions, respectively. K denotes the size of vertical analysis window.

By the similarity attribute, faults are highlighted with high values indicating significant changes in both waveform and amplitude, whereas low similarities are estimated in the continuous portions where little changes in waveform and amplitude are expected. Figure 8displays the similarity images of the horizontal slice of 1728ms and the vertical slice of inline 425, in which the faults are better imaged with higher resolution than the coherence, amplitude gradient, and semblance images (Figure 5-7).

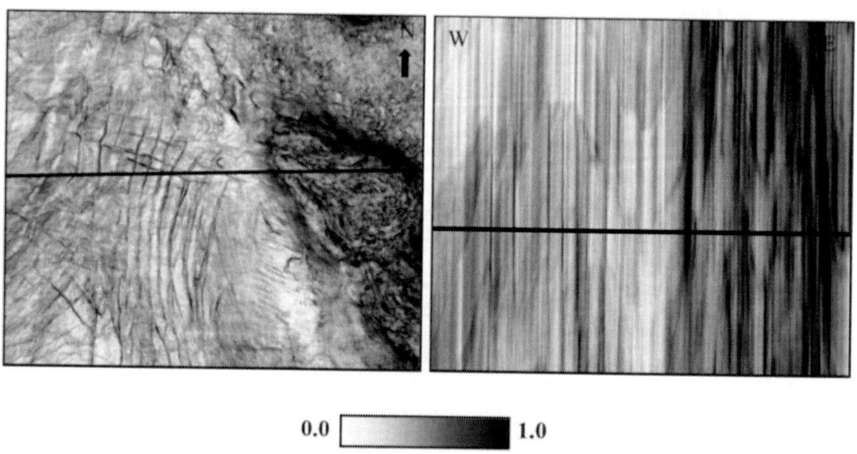

Figure 8. The horizontal slice of 1728 ms (left) and the vertical slice of inline 425 (right) of similarity.

Canny Edge Detection

As the most effective detector in the field of image processing, the Canny edge detector was introduced to seismic data (Di and Gao, 2014); and more importantly, mean removal is proposed to enhance the resolution on features where amplitude variation is often of non-zero mean. For amplitude u within a lateral analysis window of J samples and a vertical analysis window of K samples, the Canny edge detection c_5 is estimated as

$$c_5 = \frac{\sum_{k=1}^{K}\left\{\left[\sum_{j=1}^{J} f_x(j)\cdot\hat{u}(j,k)\right]^2+\left[\sum_{j=1}^{J} f_y(j)\cdot\hat{u}(j,k)\right]^2+\left[\sum_{j=1}^{J} f_x(j)\cdot\hat{u}_H(j,k)\right]^2+\left[\sum_{j=1}^{J} f_y(j)\cdot\hat{u}_H(j,k)\right]^2\right\}}{J\sum_{k=1}^{K}\sum_{j=1}^{J}\left[|f_x(j)|\cdot\hat{u}^2(j,k)+|f_y(j)|\cdot\hat{u}^2(j,k)+|f_x(j)|\cdot\hat{u}_H^2(j,k)+|f_y(j)|\cdot\hat{u}_H^2(j,k)\right]} \tag{5}$$

where $\hat{u} = u - \bar{u}$ and $\hat{u}_H = u_H - \bar{u}_H$ denote the real amplitude u and quadrature amplitude u_H after mean removal, respectively. $f_x = -\frac{x}{\sigma^2}G$ and $f_y = -\frac{y}{\sigma^2}G$ denote the Canny edge detector in inline (x-) and crossline (y-) directions, respectively. $G = \exp(-\frac{x^2+y^2}{\sigma^2})$ is the Gaussian filter, in which σ is the standard deviation.

Figure 9displays the Canny detection of the horizontal slice of 1728ms and the vertical slice of inline 425, in which an enhanced resolution is observed for faults especially in the areas with weak amplitude variation (denoted by the black circle).

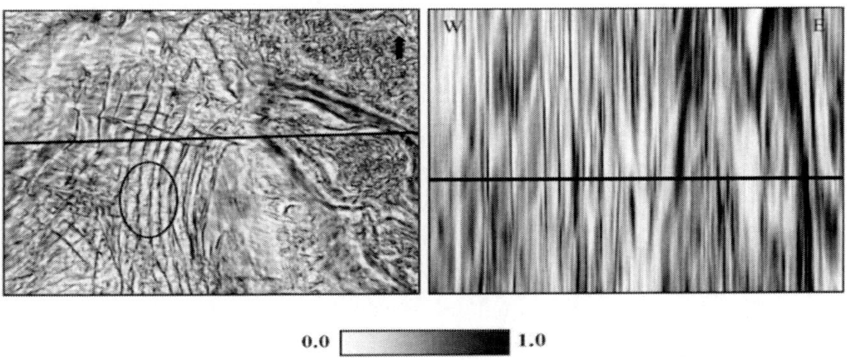

Figure 9. The horizontal slice of 1728 ms (left) and the vertical slice of inline 425 (right) of Canny edge detection.

Fault Enhancement

Even though discontinuity analysis helps identify seismic-scale faults and facilitates manual fault picking, interpretation ambiguities still exist when it is used for extracting fault patches by computers, for two major reasons. First, value of discontinuity attributes often varies laterally along a fault, which may segment the corresponding lineament into several separate parts (as shown in Figure 5-9) and thereby adds the difficulties for computers to continuously pick the fault as a whole patch. Second and more importantly, lineaments in discontinuity maps are often associated with a thickness of several pixels, especially for major faults with wide damaging zones, which adds the difficulties for computers to uniquely extract a patch that best represents fault geometry. Then fault enhancement techniques are developed to resolve both limitations, and two most popular ones are ant tracking and lineament thinning.

Ant Tracking

The ant-tracking is based on the principles of swarm intelligence and describes the collective behavior of ants finding the shortest path between the nest and a food source by communicating via pheromone, a chemical substance that attracts other ants (Pedersen et al., 2004). Figure 10 illustrates how ant tracking works with two ants starting at the same time from the nest. The ant choosing the shorter path will return back to the nest before the ant choosing the longer path, and the shorter path will be marked with more pheromone than the longer path, which attracts the next ant more likely to choose the shorter path.

When applied to seismic fault enhancement, a large number of artificial ants move along what appears to be a fault surface, emit electronic pheromone, and make decisions based on their pre-coded behavior and the pheromone along the surface. Such tracking is terminated at samples where there is no surface, only unstructured noises, or where there is a surface which does not fulfill the user-defined conditions for a fault (e.g., size, orientation). The benefits of the ant tracking are that, faults of interest could be traced by many ants at different samples and hence be strongly marked by pheromone, whereas noise and surfaces unlikely to be faults should be unmarked or weakly marked and hence be removed by thresholding.

Figure 11 displays the ant-tracking of the horizontal slice of 1728 ms and the vertical slice of inline 425. Note that the faults are highlighted more continuous and sharper (one-pixel thick). However, one major limitation of the ant tracking is the undesired overestimates of artifacts and subtle faults with small throws, which are exaggerated to the same level as the major ones (denoted by the black circle).Such exaggeration could be well avoided in manual fault picking by an experienced interpreter, but runs the risk of extracting non-fault features as patches by semi-automatic/automatic algorithms.

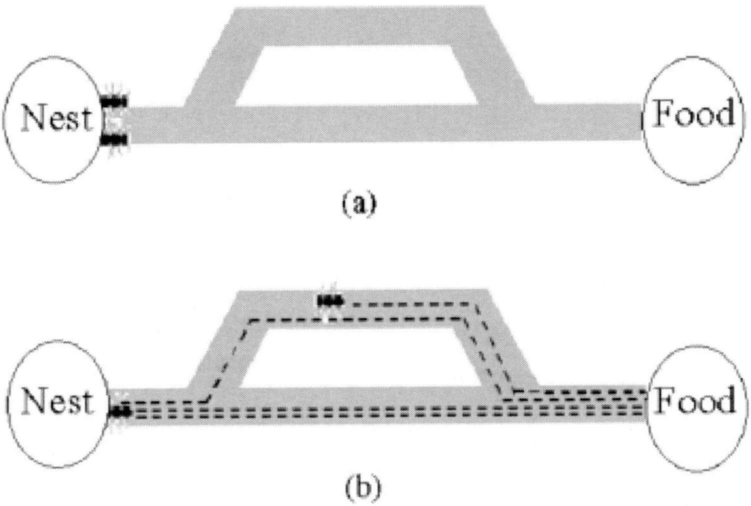

Figure 10. The sketch map of ant tracking (from Pedersen et al., 2004).

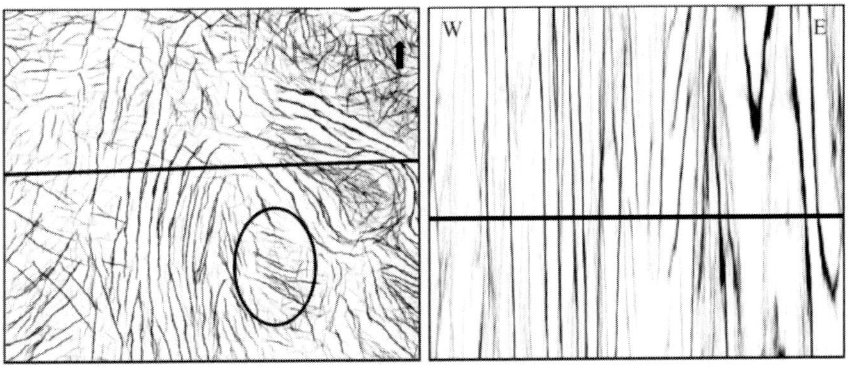

Figure 11. The horizontal slice of 1728 ms (left) and the vertical slice of inline 425 (right) of ant tracking.

Lineament Thinning

The lineament thinning works by searching for the one-pixel-thick fault surface based on the discontinuity magnitude, so that the contrast between major faults and subtle ones is well preserved without exaggerating non-fault artifacts. Various algorithms have been developed (e.g., Cohen et al., 2006; Bag & Harit, 2011; Hale, 2013; Zhang et al., 2014a), and the scanning one by Hale (2013) is considered most useful in generating the thinned fault surfaces without generating undesired bifurcation branches in the presence of noises and complex boundaries.

At each sample, the scanning algorithm first defines suitable sampling intervals of possible fault strike ϕ and dip θ that avoid undersampling. Then for each orientation (ϕ, θ), when the associated discontinuity value exceeds the maximum one stored at the sample, it is updated by the new value and also the corresponding strike ϕ and dip θ are saved. When the scanning completes, the results are images of thinned discontinuity attributes and corresponding fault strikes and dips (Hale, 2013).

Figure 12 displays the lineament thinning images of the horizontal slice of 1728 ms and the vertical slice of inline 425. It is clear that, the faults are represented by thin lineaments for interpretation, and meanwhile the artifacts and subtle lineaments in the weakly-deformed areas (denoted by the black circle) are ignored in the thinning process. Compared to the ant-tracking map (Figure 11), the thinned image serves as a better input for the semi-automatic/automatic fault patch extraction.

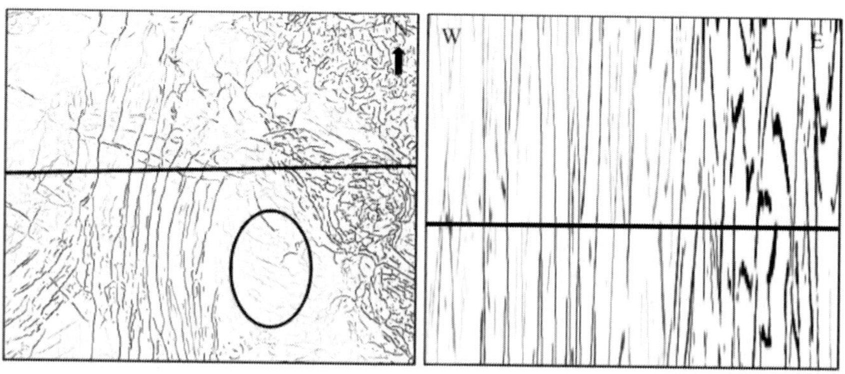

Figure 12. The horizontal slice of 1728 ms (left) and the vertical slice of inline 425 (right) of lineament thinning.

Fault Mapping

Manual Fault Picking

Traditionally, faults are interpreted by manual picking on horizontal and vertical sections of seismic amplitude, and the accuracy of manual interpretation strongly relies on the interpreter's insight and experience and conceptual understanding of the structural geology of sedimentary basins. With the recent developments in seismic attribute analysis and multi-attribute visualization, discontinuity and/or fault enhancement attributes are now often incorporated into fault interpretation by co-rendering them with the original amplitude, which helps interpreters better identify and pick the faults that are not easily discernable from amplitude due to weak waveform and/or amplitude variations. Take the North Sea dataset for example (Figure 13), manual picking is performed on a vertical section of seismic amplitude (left) and terminates in the bottom area where amplitude variation is invisible. However, when the corresponding lineament thinning is overlaid over the amplitude (right), the improved visualization allows us to continue picking the fault towards the bottom part of the vertical section, leading to more reliable interpretation of the target fault.

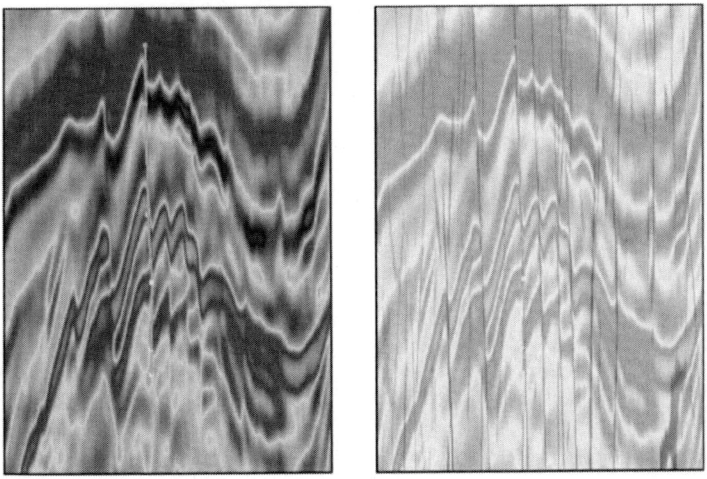

Figure 13. Manual fault interpretation (light blue) from post-stack amplitude.

Semi-Automatic Fault Extraction

Semi-automatic object detection has been successful in detecting various types of seismic features, including gas chimneys and faults. The semi-automatic fault detection often consists of three steps. First, the seismic data is scanned and a set of seismic attributes are calculated to clearly highlight faults and meanwhile reduce non-fault features. Next, faults are manually picked as seeds on a few sections of seismic amplitude and/or attributes. Finally, the manual picking is learned by a computer program and automatically extended to the entire cube, generating 3D patches throughout the whole volume.

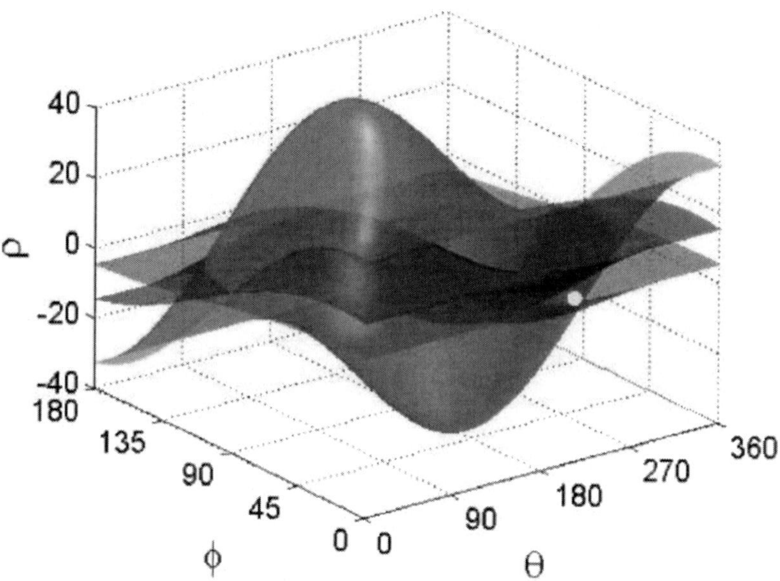

Figure 14. 3D Hough transform (from Wang & AlRegib, 2014b).

One example is the semi-automatic detection using 3D Hough transform (Wang & AlRegib, 2014b). In particular, the likely fault points are highlighted in seismic data by thresholding the corresponding discontinuity volumes. Then 3D Hough transform is applied to detect the most-likely fault planes in seismic volumes (Figure 14). Finally, after filtering the noisy planes, the weighted plane fitting method is used to extract the smoothed fault surfaces from the remaining fault planes. Figure 15 displays the faults detected in two inline sections using the semi-automatic method, demonstrating its capability of extracting patches that fit the true faults.

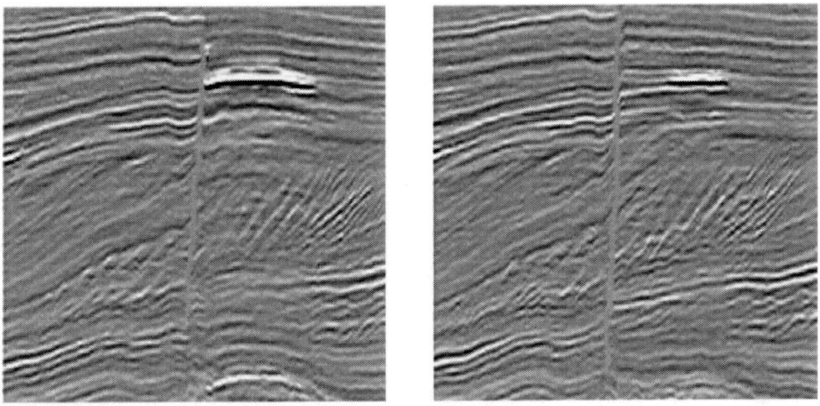

Figure 15. Semi-automatic fault extraction as shown in two seismic sections (from Wang & AlRegib, 2014b).

Automatic Fault Extraction

Compared to the manual picking and semi-automatic extraction, the method of automatic seismic object detection avoids manual picking that is time-consuming and sensitive to interpretational bias, and thus is superior in both efficiency and accuracy. The automatic fault detection also consists of three steps. First, the seismic dataset is preconditioned by filtering and/or smoothing if necessary, and discontinuity attributes are generated from the amplitude volume, which could be these mentioned above or other derivations. Then the discontinuities are thinned to be one-pixel thick by applying the fault-enhancement approaches, and the thinned lineaments allow computers to uniquely quantify the fundamental properties about a fault, such as its location, geometry and size. Finally, fault patches are extracted from the thinned lineaments, each of which represents a fault in the subsurface.

In practice, due to its sensitivity to noises present in seismic data, automatic extraction often requires complicated parameterization/optimization, among which minimum patch size and confidence level are of most importance. In particular, the minimum patch size helps eliminate small faults without interpretation interest, and a larger size is used when only the major ones are the targets of fault interpretation. The confidence level controls the sensitivity to noises and/or artificial lineaments, and a high level helps avoid extracting non-fault artifacts; however, it runs the risk of missing the true faults. Figure 16 displays the fault patches extracted from the North Sea dataset using high and low confident levels, respectively. It is clear that some small patches are not shown in the result using conservative parameters (left), part of which are artifacts but the rests may represent true faults. Therefore, even for automatic fault extraction, insights from an experienced geologist/geophysicist are always needed to resolve the conflict between high resolution and low accuracy.

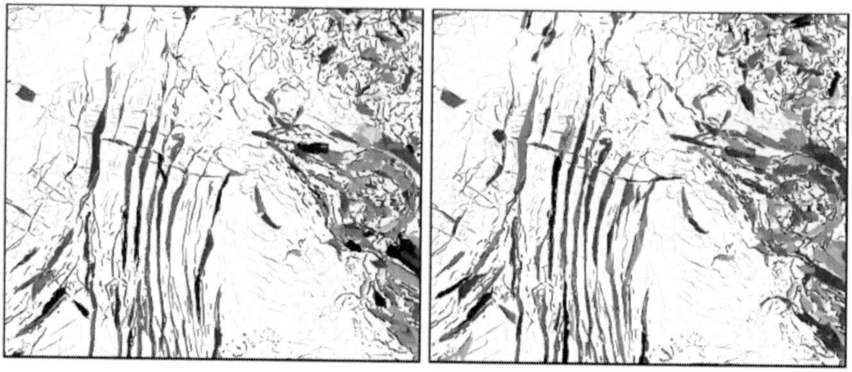

Figure 16. Automatic fault detection from high (left) and low (right) confidence level.

Fault Map Quality Control

The results from semi-automatic/automatic fault detection are often calibrated to the sections of amplitude and/or discontinuity attributes for quality control, such as validating the patch geometry and size in regular seismic images. Figure 17 and Figure 18 display the fault patches attached to the horizontal slice of 1728ms and the vertical slice of inline 320, respectively. In general, patches are successfully extracted for the major faults, and their geometry follows the actual fault geometry with high confidence. However, two major limitations are observed. First, as shown in Figure 17, a large fault is represented as several small patches instead of a large one as expected (denoted by the red circle), which could be misleading in reservoir structure interpretation and fault-network modeling. Second, as shown in Figure 18, detection of a fault often terminates before reaching the fault edges where the feature weakens to a lower level beyond the resolution of semi-automatic/automatic extraction. Therefore, the extracted patch is too small to be true, compared to its actual size and based on the interpretational experience. Both limitations could be addressed by adjusting the parameter settings according to the target fault, such as adding seed pickings in the areas where the patches stop growing for semi-automatic extraction and lowering the confidence level for automatic extraction; however, this may create new problems for picking the other faults in different orientations, decreasing the productivity and reliability for automatic fault extraction. Therefore, the automatic fault detection is still in the phase of experimental testing and not ready for reliable applications to industrial projects.

Figure 17. Fault patches calibrated to the horizontal slice of 1728 ms of lineament thinning.

Future Work

Discontinuity analysis is a fundamental process for fault enhancement and the associated patch extraction. A discontinuity attribute would help reveal and detect faults. However, existing discontinuity attributes are limited to seismic-scale faults and fractures. Beyond the discontinuity attribute, seismic geometric attributes, such as curvature and flexure (Gao, 2013; Gao and Di, 2015), can add to detect and characterize faults and fractures that fall below the seismic resolution (Di and Gao, 2016).

Fault enhancement methods, such as the lineament thinning algorithm, provides fault trace with one-pixel width, which could be implemented for improving the existing semi-automatic/automatic algorithms or developing new ones to address the conflict between detection accuracy and artifact. Take the fault autotracking algorithm by Admasu et al. (2006) for example, the thinned fault traces could greatly facilitate fault propagation, leading to enhanced fault patches in 3D space.

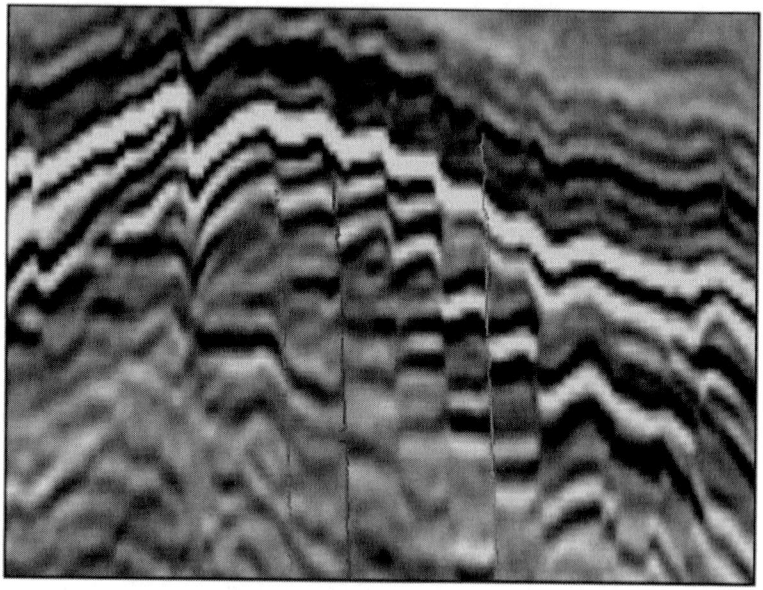

Figure 18. Fault patches calibrated to the vertical slice of inline 320 of post-stack amplitude.

Last, structure rules should always be honored when developing any geophysical algorithms for fault detection, so that the results are geologically valid. For example, a fault trace should not have abrupt bending, and thus fault patch extraction should stop where fault orientation changes dramatically. Due to the heterogeneity in finite strain along a major fault in 3D space, displacement along the fault often varies significantly, and it is common that part of it may fall below the detection criteria, leading to segmented and discontinuous fault patches. Therefore, a robust fault detection algorithm should be capable of delineating faults at various scales, identifying portions of weak faulting, and finding the associated fault gaps and tip lines in the fault propagation process.

Conclusion

In petroleum industry, there are many algorithms for fault detection from 3D seismic data. This study compares a suite of algorithms that are most helpful and applicable to discontinuity attribute analysis, fault enhancement, and automatic fault patch extraction. In general, these methods are capable of highlighting faults associated with apparent waveform/amplitude variations and extracting the corresponding fault surfaces with a fair resolution and accuracy. Problems still exist, particularly with splitting a larger fault into several small patches and detecting fault boundaries. Further efforts are needed for resolving the problems for improved fault detection from 3D seismic data.

Acknowledgments

This study is supported by the Foundation of State Key Laboratory of Shale Oil and Gas Enrichment Mechanisms and Effective Development, SINOPEC (China). Thanks go to Schlumberger and dGB Earth Sciences for software donation to the Department of Geology and Geography at West Virginia University. Data from the North Sea was downloaded from the OpendTect Open Seismic Repository (opendtect.org/osr), where it is available under a creative commons BY-SA 3.0 license.

References

Admasu, F., Back, S. & Toennies, K. (2006). Autotracking of faults on 3D seismic data. *Geophysics*, *71*, A49-A53.

Bag, S. & Harit, G. (2011). Skeletonizing character images using a modified medial axis-based strategy. *International Journal of Pattern Recognition and Artificial Intelligence, 25,* 1035-1054.

Bahorich, M. & Farmer, S. (1995). 3-D seismic discontinuity for faults and stratigraphic features: the coherence cube. *The Leading Edge, 14,* 1053-1058.

Cohen, I. & Coifman, R. R. (2002). Local discontinuity measures for 3-D seismic data. *Geophysics, 67,* 1933-1945.

Cohen, I., Coult, N. & Vassiliou, A. (2006). Detection and extraction of fault surfaces in 3D seismic data. *Geophysics, 71,* P21-P27.

de Groot, P. F. M. (1999). Seismic reservoir characterization using artificial neural networks. *19th Mintrop-Seminar, Munster, Germany.*

Di, H. & Gao, D. (2014). Gray-level transformation and Canny edge detection for 3D seismic discontinuity enhancement. *Computers & Geosciences, 72,* 192-200.

Di, H. & Gao, D.(2016). Improved estimates of seismic curvature and flexure based on 3D surface rotation in the presence of structure dip. *Geophysics*, 81, IM37-IM47.

Gao, D. (2013). Integrating 3D seismic curvature and curvature gradient attributes for fracture detection: Methodologies and Interpretational implications. *Geophysics*, 78, O21-O38.

Gao, D. & Di, H. (2015). Extreme curvature and extreme flexure analysis for fracture characterization from 3D seismic data: New analytical algorithms and geologic implications. *Geophysics, 80,* IM11-IM20.

Gersztenkorn, A., &Marfurt, K. J. (1999). Eigenstructure-based coherence computations as an aid to 3-D structural and stratigraphic mapping. *Geophysics, 64,* 1468-1497.

Gibson, D., Spann, M., Turner, J. & Wright, T. (2005). Fault surface detection in 3D seismic data. *IEEE Transactions on Geoscience and Remote Sensing, 43,* 2094-2102.

Hale, D. (2013). Methods to compute fault images, extract fault surfaces, and estimate fault throws from 3D seismic images. *Geophysics, 78,* O33-O43.

Heggland, R., Meldahl, P., Bril, A. H. & de Groot, P. F. M. (1999). The chimney cube, an example of semi-automated detection of seismic objects by directive attributes and neural networks: part II - Interpretation. *69th Annual International Meeting, SEG, Expanded Abstracts,* 935-937.

Luo, Y., Higgs, W. G. & Kowalik, W. S. (1996). Edge detection and stratigraphic analysis using 3-D seismic data. *66th Annual International Meeting, SEG, Expanded Abstracts,* 324-327.

Luo, Y., Marhoon, M., Al-Dossary, S. & Alfaraj, M. (2002). Edge-preserving smoothing and applications. *The Leading Edge, 21,* 136-141.

Meldahl, P., Heggland, R., Bril, A. H. & de Groot, P. F. M. (1999). The chimney cube, an example of semi-automated detection of seismic objects by directive attributes and neural networks: part I - Methodology. *69th Annual International Meeting, SEG, Expanded Abstracts,* 931-934.

Pedersen, S. I., Skov, T., Randen, T. & Sonneland L. Automatic fault extraction using artificial ants. In: Iske A. & Randen T. editors. Mathematical methods and modeling in hydrocarbon exploration and production: Springer; 2004; 107-116.

Tingdahl, K. M., Bril, A. H. & de Groot, P. F. M. (2001). Improving seismic chimney detection using directional attributes. *Journal of Petroleum Science and Engineering, 29,* 205-211.

Tingdahl, K. M. & de Rooij, M. (2005). Semi-automatic detection of faults in 3D seismic data. *Geophysical Prospecting, 53,* 533-542.

Wang, Z. & AlRegib, G. (2014a). Automatic fault surface detection by using 3D Hough transform. *84th Annual International Meeting, SEG, Expanded Abstracts*, 1439-1444.

Wang, Z. & AlRegib, G. (2014b). Automatic fault tracking across seismic volumes via tracking vectors. *IEEE ICIP,* 5851-5855.

Wang, Z., Temel, D. & AlRegib, G. (2014). Fault detection using color blending and color transformations. *Global SIP: Perception Inspired Multimedia Signal Processing Techniques,* 1171-1175.

Zhang, B., Liu, Y., Pelissier, M. & Hemstra, N. (2014). Semiautomated fault interpretation based on seismic attributes. *Interpretation, 2,* SA11-SA19.

Biographical Sketches

Haibin Di

Affiliation: Formerly Department of Geology and Geography, West Virginia University, Morgantown, West Virginia, US; presently Center for Energy and Geo Processing (CeGP), Georgia Institute of Technology, Atlanta, Georgia, US

Research and Professional Experience

01/2016 – 05/2016	Adjunct instructor	West Virginia Wesleyan College, US
05/2015-08/2015	Earth science intern	Chevron Energy Technology Company, US
05/2014 – 08/2014	Geophysical intern	PetroTechnical Services (Schlumberger), US
05/2013 – 08/2013	Geoscience intern	d*GB Earth Sciences*, US

Publications Last Three Years

Journal Publications (9)

- Di, H., and D. Gao, 2016, Improved estimates of seismic curvature and flexure based on 3D surface rotation in the presence of structure dip: *Geophysics*, 81, IM37-IM47.
- Di, H., and D. Gao, 2015, Volumetric extraction of most positive/negative curvature and flexure attributes for improved fracture characterization from 3D seismic data: *Geophysical Prospecting*, doi: 10.1111/1365-2478.12350.
- Gao, D., and H. Di, 2015, Extreme curvature and extreme flexure analysis for fracture characterization from 3D seismic data: New analytical workflows and geologic implications: *Geophysics*, 80, IM11-IM20.
- Di, H., and D. Gao, 2014, Gray-level transformation and Canny edge detection for 3D seismic discontinuity enhancement: *Computer & Geosciences*, 72, 192-200.
- Di, H., and D. Gao, 2014, A new algorithm for evaluating 3D curvature and curvature gradient for improved fracture detection: *Computer & Geosciences*, 70, 15-25.
- Zheng, Z., P. Kavousi, and H. Di, 2014, Multi-attribute and neural network-based fault detection in 3D seismic interpretation: *Advanced Materials Research,* 838-841, 1497-1502.
- Di, H., and L. Xu, 2012, Calculation of the ground motion generated by a finite source in stratified media: An extension of the point-source case: *Acta Seismologica Sinica* (in Chinese), 34(4), 425-438.
- Di, H., Y. Guo, and X.W. Liu, 2011, The inversion algorithm of seismic relative acoustic impedance based on Cauchy sparseness constraint Bayesian estimation: *Geophysical Prospecting For Petroleum* (in Chinese), 50(2), 124-128.
- Xu, L., H. Di, W. Feng, et al, 2010, Estimation of the fault-near ground motion of the 2010 Yushu, Qinghai, Ms7.1 earthquake: *Chinese J. Geophys.* (in Chinese), 53(6),1366-1373.

Conference Presentations (8)

[1] Di, H., and D. Gao, 2016, Improved seismic texture analysis based on non-linear gray-level transformation: 86th Annual Meeting, SEG, Technical Program Expanded Abstracts.

[2] Di, H., and D. Gao, 2015, Reflection geometry-based strain analysis from 3d seismic data: 85th Annual Meeting, SEG, Technical Program Expanded Abstracts, 1922-1926.

[3] Di, H., and D. Gao, 2015, A new analytical method to evaluate maximum flexure for improved fracture characterization from 3-D seismic data: AAPG Annual Convention and Exhibition, Search and Discovery Article #90216.

[4] Gao, D., and H. Di, 2015, Texture inversion from seismic amplitude: Application to facies characterization, offshore Angola, West Africa: AAPG Annual Convention and Exhibition, Search and Discovery Article #90216.

[5] Di, H., and D. Gao, 2014, An algorithm for evaluating Fresnel-zone textural roughness for seismic facies interpretation: American Geophysical Union (AGU) Fall Meeting, NS43A-3857

[6] Di, H., and D. Gao, 2014, A new analytical method for azimuthal curvature analysis from 3D seismic data: 84th Annual Meeting, SEG, Technical Program Expanded Abstracts, 1634-1638.

[7] Di, H., and D. Gao, 2014, Predicting fracture orientations with volumetric curvature gradient analysis: case study from Teapot Dome in Wyoming: AAPG Annual Convention and Exhibition, Search and Discovery Article #41331

[8] Di, H., and D. Gao, 2013, Gray-level transformation and Canny edge detection for 3D seismic discontinuity enhancement: 83'd Annual Meeting, SEG, Technical Program Expanded Abstracts, 1504-1508.

Dengliang Gao

Affiliation: Department of Geology and Geography, West Virginia University, Morgantown, West Virginia, US

Research and Professional Experience

08/2015– Present	Professor	West Virginia University, US
08/2009-08/2015	Associate Professor	West Virginia University, US
04/2008 – 03/2009	Staff Geophysicist	Chevron Energy Technology Company, USA
05/1998 – 04/2008	Senior Geologist	Marathon Oil Company, US
05/1997 – 05/1998	Post Doc Fellow	Exxon Production Research Company, USA

Publications Last Three Years:

[1] Gao, D., 2016, Comments on "Evidence of Recent Thrust Faulting on the Moon Revealed by the Lunar Reconnaissance Orbiter Camera": e-Letters of *Science.*

[2] Donahoe, T., and D. Gao, 2016, Application of 3D seismic attribute analysis to structure interpretation and hydrocarbon exploration southwest Pennsylvania, Central Appalachian Basin: A case study: *Interpretation*, 4(3), T291-T302.

[3] Di, H., and D. Gao, 2016, Improved estimates of seismic curvature and flexure based on 3D surface rotation in the presence of structure dip, *Geophysics*, 81, IM37-IM47.

[4] Di, H., and D. Gao, 2015, Volumetric extraction of most positive/negative curvature and flexure attributes for improved fracture characterization from 3D seismic data: *Geophysical Prospecting*, doi: 10.1111/1365-2478.12350.

[5] Gao, D., and H. Di, 2015, Extreme curvature and extreme flexure analysis from 3D seismic data: Implications for fractured reservoir characterization: *Geophysics*, 80, DOI: 10.1190/GEO2014-0185.1.

[6] Di, H., and D. Gao, 2014b, Gray-level transformation and canny edge detection for 3D seismic discontinuity enhancement: *Computers & Geosciences*, 72, 192-200.

[7] Di, H., and Gao, D., 2014a, A new algorithm for evaluating 3D curvature and curvature gradient for improved fracture detection: *Computers & Geosciences*, 70, 15-25.

[8] Gao, D., 2013b, Wavelet spectral probe for seismic structure interpretation and fracture characterization: A workflow with case studies: *Geophysics*, 78, O57–O67.

[9] Gao, D., 2013a, Integrating 3D seismic curvature and curvature gradient attributes for fracture characterization: Methodologies and interpretational implications: *Geophysics*, 78, O21-O31.

In: Fault Detection
Editor: Daniel Martin

ISBN: 978-1-53610-345-8
© 2017 Nova Science Publishers, Inc.

Chapter 4

SELF-ADAPTIVE EXPERT SYSTEM FOR PROCESS MONITORING AND FAULT DETECTION

Milena R. Petković**, *Željko S. Kanović* and *Milan R. Rapaić
Faculty of Technical Sciences, Novi Sad, Serbia

Abstract

Process monitoring and early fault detection are paramount in modern industries. A well-developed process monitoring system is expected to reduce failure rates, production costs, and prevent health and environmental hazards. An abundance of sophisticated monitoring and fault detection strategies is available, most of which require either accurate process models, including models of the most frequently occurring faults, or large databases filled with relevant historical data. Unfortunately, neither of the previous is usually available, and process monitoring is often reduced to more-or-less elaborate limit-checking. Even in the case when the models and/or historical datasets are available, the obtained monitoring techniques are highly specialized and must usually be re-developed for any new application at hand. This chapter presents a general-purpose, self-adaptive expert system for online and offline process monitoring with possibility of fault detection, capable to learn the properties of the process being monitored and to identify and separate its working regimes, both nominal and unwanted ones. The proposed system is also capable to identify changes in the process

* E-mail address: milena5@uns.ac.rs (Corresponding author).

operation; both abrupt and incipient ones. The flexibility of the system is achieved by separating the monitoring algorithm into two distinct layers. The first layer, which is application dependent, is called Feature Generator (FG). FG uses current measurements of process variables to compute "features," numerical values used to concisely describe current working conditions. Features are then fed to the second layer, Potential Based Recursive Clustering (PBRC) algorithm, an original, novel adaptive clustering procedure. In essence, PBRC is designed as self-learning and auto-growing neural-network which tracks, in a robust manner, the motion of the process in the features space. It identifies, in an adaptive fashion, areas corresponding to separate process working conditions and decides which of the working conditions is currently active. If a novel working regime is encountered, PBRC can enlarge its knowledge base and incorporate the novelty for further reference. The system was tested and validated on a number of laboratory installations and also in real industrial environment. Obtained results showed that proposed system can detect faults and changes in working regime in the early stages with high accuracy and reliability.

Keywords: fault detection and diagnosis, expert systems, characteristic features, preventive maintenance, intelligent monitori

1. Introduction

Complex plants are disposed to faults. Even though good design intends to minimize the occurrence of faults, recognition that faults do happen enables system operators to respond appropriately so that the effects of the faults are minimized. Fault is any change in a plant that prevents it from operating correctly. Fast, robust and reliable fault detection is an important part in a successful maximization of energy efficiency and productivity. In many cases it is also very important to determine exactly what kind of fault happen. This procedure is called fault detection and isolation (FDI). Systems for fault detection and isolation in plants detect and try to reimburse faults in one or more of the following three subsystems: the actuator, the plant and sensors. There are three key performance criteria for fault detection and isolation systems- false alarm rate, missed alarm rate and time delay. The design of systems for fault detection and isolation demands the careful consideration of these criteria and tradeoffs. Expert systems are very suitable for fault detection because they are easy to develop, have capability to reason under uncertainty and the capability to deliver explanations for the solutions provided. The main advantages in the development of expert systems for diagnostic problem solving are: ease of development, transparent reasoning, and the ability to reason under uncertainty and the ability to provide explanations for the solutions provided. There are a number of researchers who have worked on the application of expert systems for FDI problems (Becraft et al., 1993; Kulić et al., 2011; Kanović et al., 2013; Matić et al., 2010; Tarifa et al., 1997).

Number of classification and clustering techniques has been applied successfully in various fields of science and engineering. A general overview of different statistical procedures can be found in (Izenman, 2008), while a gamut of soft computing techniques, including techniques based on artificial neural networks, support vector machines (SVMs) and fuzzy logic, is presented in (Kecman, 2001). Most of the proposed techniques are capable of distinguishing among several fixed and a priori known classes (cluster centers). There are however cases in which all classes are not known in advance, and also cases where the properties of the existing classes change (evolve) over time. In such cases it would be preferable to use a classification algorithm which is capable of learning from the input data stream and evolving.

Fault detection and isolation (FDI) is an important filed of modern process engineering. A survey of modern approaches to fault diagnosis can be found in (Venkatasubramanian et al., 2003a; Venkatasubramanian et al., 2003b, Venkatasubramanian et al., 2003c). It has been claimed (Venkatasubramanian et al., 2003a) that any fault diagnosis technique can be seen as (or at least reduce to) a classification procedure: a fault is either present or it is not. A more sophisticated diagnosis classifier would even be capable of distinguishing among different types of faults, and thus perform fault isolation. However, it is rather difficult to develop a diagnosis procedure which would be capable of identifying novel faulty conditions, ones not seen during design phase. Also, most FDI techniques are incapable of coping with the drift of some of the existing working conditions. The ability to identify novelties and adapt to the changing environmental conditions is identified as one of the desirable features of any fault diagnosis technique (Venkatasubramanian et al., 2003a).

The aim of the present chapter is to present an adaptive clustering procedure capable of identifying novelties and respond to changes in the input data stream designed for fault detection. When applied in an FDI setting, the proposed procedure should be capable of detecting working conditions which occur for the first time during the exploitation of the algorithm. These newly recognized working conditions can be labeled as either nominal or faulty by an experienced operator. The proposed technique can be seen as an evolving classifier and is inspired by the on-line identification procedure proposed by Angelov and Filev in (Angelov et al., 2004) as well with the eClass family of evolving fuzzy classifiers (Angelov et al., 2004). The proposed algorithm was originally presented in (Petkovic et al., 2012) and in this chapter we will use the term Potential Based Recursive Clustering Algorithm (PBRC). The proposed technique may be seen as an alternative to the classical model-based fault detection techniques (Ding, 2008) and also to hybrid fault detection techniques, such as the one presented in (Rapaic et al., 2010).

The chapter is organized as follows. The key underlying notion of information potential is introduced in Section 2. The information potential used here is different than the information potential utilized in (Angelov et al., 2004).The general structure of the proposed expert system based on adaptive potential based recursive clustering procedure is presented in Section 3. Specialization of the proposed algorithm for FDI purposes is described in Section 4.

Results of system testing performed oil industry facilities are presented are presented in Section 5.

2. Informational Potential

In (Angelov et al., 2004) Angelov and Filev proposed an adaptive clustering procedure relying on the notion of the information potential. An information potential of a point with respect to an ordered set of points is a measure of similarity between the point and the set. If the information potential of a point is high, the point is representative with respect to other points within the set. Points with the highest values of the information potential are candidates for cluster centers.

Let us define the notion of the information potential more formally. Consider a set of *n features*, each of which is represented as a real number. The feature space $\mathcal{F}$ is therefore a subset of $\mathbb{R}^n$. Let $\mathbf{z} \in \mathbb{R}^n$ be an arbitrary feature vector and let $\mathcal{Z} = \{\mathbf{z}_j \mid j = 1 \dots N\} \subseteq \mathcal{F}$ be an ordered set of known feature vectors. The assumption is that the classification is made online, recursively as each new feature point becomes available. Elements of the feature set are ordered in the same sequence they become available: $\mathbf{z}_0$ is the first obtained feature vector, $\mathbf{z}_N$ is the most recent one. In order to emphasize the generality of the proposed classifier, the feature vectors will be denoted simply as the *data points*.

The **mean square distance** of a data point **z**with respect to the ordered data set $\mathcal{Z}$ is defined as (Angelov et al., 2004)

$$S(\mathbf{z}, \mathcal{Z}) = \frac{1}{1+N} \sum_{j=0}^{N} \left\| \mathbf{z} - \mathbf{z}_j \right\|_2^2 \tag{1}$$

where $\|\mathbf{q}\|_2^2 = \mathbf{q}^T\mathbf{q}$ denotes the standard Euclidean 2-norm, and *T*in superscript denotes matrix- (or in this case, vector-) transposition. As usual, all vectors are implicitly assumed to be column-vectors.

Angelov and Filev defined the information potential as a Cauchy-type function of the mean square difference

$$P(\mathbf{z}, \mathcal{Z}) = \frac{1}{1 + S(\mathbf{z}, \mathcal{Z})} \tag{2}$$

Information potentional defined in such manner represents meashure od similarity between point **z** and ordered data set $\mathcal{Z}$.

The important property of the information potential is that its value cannot change abruptly. The information potential is computed with respect to all previously seen data points, and thus if the underlying data shift in the feature space, it wouldtake some time for the value of the information potential to reflect this change. Normally this is good and desirable, since it is the basic mechanism providing resilience to measurement noise. Any outliers, invalid and noisy data points, would typically be isolated and thus would not have the capability of significantly changing potential values. However, in the fault diagnosis setting, the fact that all previous data points are taken into consideration may impose a significant drawback. It is the nature of faults to appear after long periods of nominalprocess operations. Depending on the sampling rate, after some operation time the nominal data points in $\mathcal{Z}$ will vastly outnumber faulty ones. Consequently, the ability of the average square distance (and thus also the information potential) to detect changes in the feature data stream will be impaired.

To overcome this problem, the notion of mean square distance will be redefined. Let us define the **exponentially windowed mean square distance** between an arbitrary data point **z** and all data points belonging to $\mathcal{Z}$ as

$$S_\lambda(\mathbf{z}, \mathcal{Z}) = (1-\lambda)\sum_{j=1}^{N} \lambda^{N-j} \left\|\mathbf{z} - \mathbf{z}_j\right\|_{\mathbf{w}}^2 \tag{3}$$

where $\|\cdot\|$ may, in principle, denote an arbitrary norm in the feature space. In the sequel, however, $\|\cdot\|$ will always denote the **weighted Euclidean 2-norm**, defined as

$$\|\mathbf{q}\|_{\mathbf{W}}^{2} = \mathbf{q}^{T}\mathbf{W}\mathbf{q} \tag{4}$$

with **W** being a suitably chosen, positive definite, symmetric matrix. Practically, the weight matrix is usually diagonal, and serves the purpose of data scaling. Thus, the overall mean square distance (1) is replaced by the mean square distance (4) which effectively takes into consideration the recent points only. In fact, only the points appearing within an exponential time window are accounted for. The parameter λ controls the width of this window and has the same purpose as the forgetting factor in the recursive least squares algorithm. The λ parameter should be chosen in range [0,1]. Depending on the sample rate, the noisiness of the signal and the dynamics of the process, the usual choice is between 0.9 and 1. The information potential can be defined as a Cauchy function of the newly introduced exponentially weighted mean square distance (3). This is the approach proposed in (Angelov et al., 2004). It is however also possible to use other, more general definitions.

Let us define the potential of a data point **z**with respect to the data set $\mathcal{Z}$ as a positive, real-valued mapping of the exponentially weighted mean square distance, $P(\mathbf{z},\mathcal{Z},\lambda) \equiv P(S(\mathbf{z},\mathcal{Z},\lambda))$, with the following properties: for all $S \geq 0$,

- (P1) $P(S) \in (0,1]$
- (P2) $P(S) = 1 \Leftrightarrow S = 0$
- (P3) $P(S) = 0 \Leftrightarrow S \to \infty$
- (P4) P is a differentiable, monotonically decreasing function of S, $P'(S) < 0$.

The first property (P1) simply states that a potential takes values from a compact set [0, 1].

The property (P2) states that a point **z**has the potential equal to 1 if and only if all the points in Z are equal to **z**.

(P3) states that the potential becomes equal to zero if and only if the distance to at least some point in Z grows without an upper bound; a situation which is not to be expected in practice.

(P4) implies that there is a well-defined inverse function $P^{-1}:[0,1]\rightarrow[0,\infty]$. It is easily seen that the Cauchy-type function (3) satisfies requirements (P1)-(P4).

Any functions satisfying properties (P1)-(P4) can be considered as a mean proximity measure. In fact, the clustering procedure can be designed solely on the basis of the mean square distance, but it is more convenient to use the notion ofthe information potential. By using the information potential, the proposed algorithm can be tightly related to the entire family of fuzzy classifiers, as well as to the evolving eClass classifiers (Kecman, 2001; Angelov et al., 2004).

An important property of the previously introduced average square distance and information potential is that it is possibly to compute them recursively. In other words, any classification algorithm which is based on them can be implemented in real time. This is shown by the following two theorems. The following conventions be adopted in the proof: $P(\mathbf{z},\mathcal{Z}_k,\lambda)\equiv P_k(\mathbf{z})$, $S(\mathbf{z},\mathcal{Z}_k,\lambda)$ $\equiv S_k(\mathbf{z})$.

Theorem 1. Let $S(\mathbf{z},\mathcal{Z}_k,\lambda)\equiv S_k(\mathbf{z}_k)\equiv S_k(\mathbf{z})$ be defined by (4), and let the norm be defined as in (5). The following recursion holds

$$S_k=\lambda S_{k-1}+2\lambda(1-\lambda)(\mathbf{z}_k-\mathbf{z}_{k-1})^T\,\mathbf{W}\mathbf{F}_k+\lambda(1-\lambda)\|\mathbf{z}_k-\mathbf{z}_{k-1}\|_{\mathbf{W}}^2\,Z_k \tag{6}$$

$$S_0=0 \tag{7}$$

$$\mathbf{F}_k=\lambda\mathbf{F}_{k-1}+\lambda Z_{k-1}(\mathbf{z}_{k-1}-\mathbf{z}_{k-2}) \tag{8}$$

$$\mathbf{F}_0=0 \tag{9}$$

$$Z_k=\lambda Z_{k-1}+1 \tag{10}$$

$$Z_0=0 \tag{11}$$

Proof. The key point is to note that

$$\begin{aligned}\|\mathbf{z}_k-\mathbf{z}_i\|_{\mathbf{W}}^2 &= (\mathbf{z}_k-\mathbf{z}_i)^T\mathbf{W}(\mathbf{z}_k-\mathbf{z}_i)\\ &= (\mathbf{z}_k-\mathbf{z}_{k-1}+\mathbf{z}_{k-1}-\mathbf{z}_i)^T\mathbf{W}(\mathbf{z}_k+\mathbf{z}_{k-1}-\mathbf{z}_{k-1}-\mathbf{z}_i)\\ &= (\mathbf{z}_k-\mathbf{z}_{k-1})^T\mathbf{W}(\mathbf{z}_k-\mathbf{z}_{k-1})+2(\mathbf{z}_k-\mathbf{z}_{k-1})^T\mathbf{W}(\mathbf{z}_{k-1}-\mathbf{z}_i)\\ &\quad+(\mathbf{z}_{k-1}-\mathbf{z}_i)^T\mathbf{W}(\mathbf{z}_{k-1}-\mathbf{z}_i)\\ &= \|\mathbf{z}_k-\mathbf{z}_{k-1}\|_{\mathbf{W}}^2+2(\mathbf{z}_k-\mathbf{z}_{k-1})^T\mathbf{W}(\mathbf{z}_{k-1}-\mathbf{z}_i)+\|\mathbf{z}_{k-1}-\mathbf{z}_i\|_{\mathbf{W}}^2\end{aligned} \tag{12}$$

By means of the above expression, it is readily obtained that

$$\begin{aligned}S_k &= (1-\lambda)\sum_{i=0}^{k}\lambda^{k-i}\|\mathbf{z}_k-\mathbf{z}_i\|_{\mathbf{W}}^2 = (1-\lambda)\sum_{i=0}^{k-1}\lambda^{k-i}\|\mathbf{z}_k-\mathbf{z}_i\|_{\mathbf{W}}^2\\ &= (1-\lambda)\sum_{i=0}^{k-1}\lambda^{k-i}\|\mathbf{z}_k-\mathbf{z}_{k-1}\|_{\mathbf{W}}^2+2(1-\lambda)\sum_{i=0}^{k-1}\lambda^{k-i}(\mathbf{z}_k-\mathbf{z}_{k-1})^T\mathbf{W}(\mathbf{z}_{k-1}-\mathbf{z}_i)\\ &\quad+(1-\lambda)\sum_{i=0}^{k-1}\lambda^{k-i}\|\mathbf{z}_{k-1}-\mathbf{z}_i\|_{\mathbf{W}}^2\end{aligned} \tag{13}$$

By defining

$$Z_k=\sum_{i=0}^{k-1}\lambda^{k-i},\ \mathbf{F}_k=\sum_{i=0}^{k-1}\lambda^{k-i}(\mathbf{z}_{k-1}-\mathbf{z}_i)=\sum_{i=0}^{k-2}\lambda^{k-i}(\mathbf{z}_{k-1}-\mathbf{z}_i) \tag{14}$$

the recursive expressions (6), (8) and (10) are obtained using simple algebraic manipulations.

Theorem 2. Let$S(\mathbf{z}_i^*, \mathcal{Z}_k, \lambda) \equiv S_k(\mathbf{z}_i^*)$be defined by (4), andlet the norm be defined by (5). Then

$$S_k(\mathbf{z}_i^*) = (1-\lambda)\|\mathbf{z}_k-\mathbf{z}_i^*\| + \lambda S_{k-1}(\mathbf{z}_i^*) \tag{15}$$

Proof. By definition,

$$S_k(\mathbf{z}_i^*) = (1-\lambda)\sum_{j=0}^{k}\lambda^{k-j}\left\|\mathbf{z}_i^*-\mathbf{z}_j\right\|^2 \tag{16}$$

The recursive form is obtained by extracting the most recent term in the sum

$$S_k(\mathbf{z}_i^*) = (1-\lambda)\sum_{j=0}^{k-1}\lambda^{k-j}\left\|\mathbf{z}_i^* - \mathbf{z}_j\right\|^2 + (1-\lambda)\|\mathbf{z}_i^* - \mathbf{z}_k\|^2$$
$$= \lambda(1-\lambda)\sum_{j=0}^{k-1}\lambda^{k-1-j}\left\|\mathbf{z}_i^* - \mathbf{z}_j\right\|^2 + (1-\lambda)\|\mathbf{z}_i^* - \mathbf{z}_k\|^2 \quad (17)$$

which is equivalent to (12).

The two theorems above testify that it is possible to implement a recursive clustering procedure based on an arbitrary potential function provided it depends solely on the exponentially windowed mean square distance (4). The Theorem 1 shows how to compute the potential of the new feature vector, while the Theorem 2 shows how to recursively update the potential of the cluster centers.

3. Potential Based Recursive Clustering Algorithm

The information potential defined in the previous section reflects a measure of similarity between the current and recent process behavior. If the information potential of the feature point associated with the current process behavior is high, then there has not been any sudden change in the process behavior. If, however, the potential of the current feature is low, then the abrupt change in process behavior is likely to have happened in recent past. The algorithm itself keeps track of a certain number of data points (features) which previously had high values of the information potential. These features will be referred to as **focal pointsor** or **foci**, and sometimes also as **nodes**. Each node is, in fact, a feature vector found to be the most representative for certain area of the feature space (and thus for certain working condition in the FDI setting). The set of all nodes defines the underlying **knowledge baseof** the algorithm. Information potential is computed recursively for each of the existing nodes using Theorem 2. The **active node (activefocus)** is the one with the highest value of the information potential. The **estimated current working condition** *is the condition associated with the active node.* Thus, theclassification is performed in a *winner-takes-all* manner.In order to provide adaptability the information potential isalso computed for the most recently observed feature point (the **current feature**). If this information potential is high, *higher than the potential of any of the existing feature points*,the current feature vector is more representative of the currentworking regime than any of the focal points present: *the knowledge base of the algorithm should be modified.* Themodifications can be twofold. If some of the previously seenclusters changed slightly, or if a better representative of anexisting cluster is found, the position of some focal pointschanges, but the knowledge base remain of the same size. If, however, a new cluster (potentially, a new working condition)is found, the knowledge base grows. More precisely, if thecurrent working condition is similar to some of the existingnodes, the existing node is replaced by the current featurevector. If, on the other hand, the current feature is far from any of the existing nodes, the current feature is proclaimed as the new focal point and the underlying data structure of the algorithm grows. The pseudo-code of the algorithm is presented in Listing 1.

1. Choose an appropriate feature generator.
2. Choose algorithm parameters.
3. Initiate the knowledge base
4. **BEGIN LOOP**
5. Obtain current feature vector
6. Compute, on the basis of Theorem 1, the information potential of the current feature vector.
7. Compute, on the basis of Theorem 2, the information potential of the existing nodes.
8. **IF** the information potential of the current feature vector is higher than the information potential of all existing nodes **THEN**
9. **IF** the current feature vector is close to some of the existing nodes **THEN**
10. Modify the knowledge base: replace the closest existing node with the current feature vector
11. **ELSE**
12. Extend the knowledge base: create a new node equal to the current feature vector.

END IF
END IF
END LOOP

Listing 1. Pseudo-Code of the Proposed Potention Based Recursive Clustering Procedure (PBRC)

The algorithm may be initialized on the basis of *a priori* knowledge. In this case, the initial nodes are selected as theexpected positions of typical cluster centers in the featurespace. In the FDI setting, the initial nodes should be selectedso to correspond to nominal working conditions and also to thetypical faults. If, however, a priori knowledge is not available,the algorithm may be left uninitialized. In such a case, the firstfeature vector is selected as the initial node.

In any time instant, the number of working conditions thealgorithm may identify is equal to the number of nodes. Each node is representative for a set of feature values, i.e., each node is representative for a specific working condition of theunderlying process. The proposed algorithm, or better say its knowledge base,may be seen as an self-adaptive artificial neural network (ANN) with structure presented in Figure 1. The hidden layer of the network contains the nodes of the algorithm. Once a new measurement is available, the existing nodes are compared to it, similar as with the radial basis function neural networks (RBF-NN). The difference is that the information potential is used as the basis for comparison, not the algebraic distance. The advantage of using the information potential is in theconsiderable robustness to noise and outliers (Angelov et al., 2004). Apart from that, the neural network is constantly being trained, both in the terms of its parameters (positions of the nodes in the hidden layer) as well as in terms of its structure, since the number of nodes is growing.

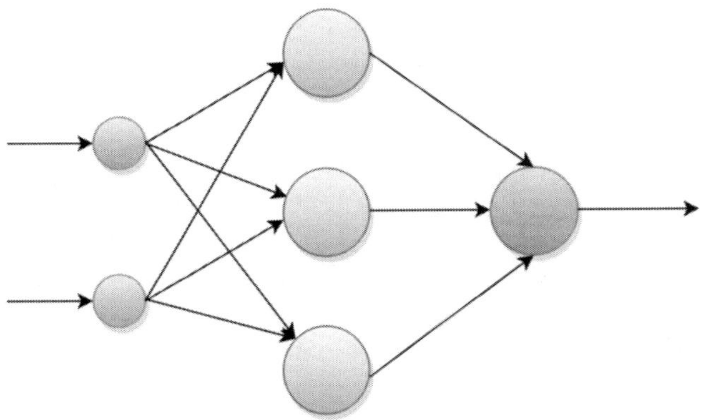

Figure 1. The general, network-like structure of the knowledge base.

4. Specialization of the PBRC Algorithm for FDI

When applying the proposed adaptive clustering procedure in the field of fault detection it is important to select a **feature generator** properly. A feature generator is an algorithm, or a device, which extracts features from process measurements. In this paper, we consider faults which change the dynamic behavior of the process under consideration. Thus, it is natural to choose a *Kalman filter* (Astrom et al., 1997.) as a feature generator. Once the dynamic behavior of the process changes, so will the estimated values of its model parameters. The proposed adaptive classification algorithms will search for the points in the parameter space which are most representative for current working regimes. If the underlying process is nonlinear, which is usually the case, a single working conditions may be described by multiple linear models, i.e., by several focal points in the feature space. It is impossible for the detection algorithm to automaticallydetect that several focal points are in fact related to the same working condition. This should be done by an experienced operator. The operator should also be responsible for accurately describing each of the focal points, or at least as labeling them as nominal or faulty. Even more, a detailed description may be assigned to each of the recognized working conditions. In the case of faulty working regimes, the assigned description may even contain a prescription of actions needed to either fix the faulty conditions or minimize its impact. Thus,the proposed algorithm is most effective when suitably combined with expert knowledge. The overall structure of the proposed expert fault diagnosis system is presented in Figure 2.

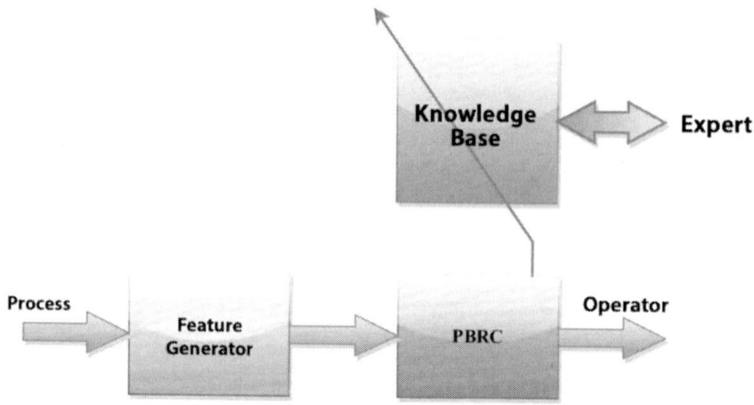

Figure 2. The general structure of the proposed adaptive FDI scheme.

5. System Testing and Results

In the first study expert system based on PBRC algorithm for simultaneous grouping and sorting the data has been successfully applied for detection ofbroken bar in squirrel-cage induction motors. The procedure is tested in company VICTORIA OIL from Sid, Serbia. We analyzed the vibration signals recorded in the facilities of the factory Victoria Oil on cage asynchronous motor.

The procedure consists of the following stages:

- Data collecting
- Signal processing and characteristic features extraction
- Detection and classification of faults, operating conditions and regimes

Each of these phases will be shown in detail below.

Data Collecting

As part of this phase, collecting measured data from the sensor vibration is done. We used accelerometers as sensors for vibration measurement, which registered acceleration for components on which they were attached and turned it into an electrical signal. This signal was standardized by means of the transmitter and used for further processing.

Signal Processing and Extraction of Characteristic Features

Vibration signals which were collected from sensors represent acceleration of the components as a series of numerical values $x(i)$ in the time domain. Such representation of the signal allows the calculation of certain statistical parameters (features) that can serve as indicators of the general state of the induction motor and the presence of certain faults.

As one of the most used features in the time domain in the literature we will mention the effective value of the signal i.e., Root Mean Square (RMS), defined by the expression:

$$x_{rms} = \left(\frac{1}{N} \sum_{i=1}^{N} x^2(i) \right)^{1/2} \tag{18}$$

This characteristic reflects a measure of "strength" of the vibration signal and can be used as an indicator of an imbalance of engine components or bearings and rotor damage. Due to the precise fault diagnosis, this feature is usually combined with the so-called "Crest" factor, which is defined by the equation

$$C = \frac{\max(x(i)) - \min(x(i))}{x_{rms}} \tag{19}$$

and represents the ratio of signal range and the effective value. This feature is an indicator of the appearance of high peak value signal or pulse vibrations, which may indicate some damage of motor bearings (Stepanić et al., 2009).

In addition to these parameters, some individual faults can be detected using the analysis of the probability density function of vibration signals. It is noted that, in proper induction motors, probability density has a Gaussian distribution, while in the presence of faults on bearings and other rotating elements, it has mostly a non-Gaussian distribution. Therefore, the following parameters characterizing the probability density function are used as the characteristic features in the time domain:

- Mean $\bar{x} = \frac{1}{N}\sum_{i=1}^{N} x(i)$
- Square mean root: $x_r = \left(\frac{1}{N}\sum_{i=1}^{N} |x(i)|^{1/2}\right)^2$
- Skewness $x_{ske} = \frac{1}{N}\sum_{i=1}^{N} (x(i)-\bar{x})^2$
- Kurtosis $x_{kur} = \frac{1}{x_v^{\,2}}\left(\frac{1}{N}\sum_{i=1}^{N} (x(i)-\bar{x})^4\right)$, where x_v is the variance of x, $x_v = \frac{1}{N}\sum_{i=1}^{N} \left(x(i)-\bar{x}\right)^2$
- Shape factor: $S = \frac{x_{rms}}{\bar{x}}$
- L factor: $L = \frac{\max\left(x(i)\right)-\min\left(x(i)\right)}{x_r}$
- I factor $I = \frac{\max\left(x(i)\right)-\min\left(x(i)\right)}{\bar{x}}$

These time domain features are not sufficient to reliably determine the origin and type of failure, because they do not provide unambiguous information. Therefore, together with the features in time domain, some characteristics in the frequency domain are also used. Since each type of failure causes an increase in the signal amplitude of vibration at very specific frequencies, spectral analysis can reliably determine the origin and type of failure. For this analysis, usually a fast Fourier transform (FFT) is used to calculate the spectrum of the vibration signal.

The most common faults of the rotor are one or more broken bars and eccentricity of the gap between the rotor and stator. Damage of rotor bars occurs due to thermal stresses, fatigue or effects of stress during transient operation, which is especially frequent in large engines. The damaged rotor bar causes a significant reduction in engine torque and very adversely affects the operation of the engine. Damaged rotor bar is recognized in the spectrum of the vibration signal as the presence of increased amplitude at sideband frequencies, given by expression

$$f_{brb} = (1 \pm 2ks) f_r \,, \quad k = 1, 2, ... \tag{20}$$

where f_{brb} is frequency of rotation of the rotor, and s represents slip. Of particular interest are first order sidebands (k = 1), and the amplitudes of signal spectrum at these frequencies are usually taken as the characteristic features for detection of this fault. Left sideband $(1-2s)f_r$ occurs due to electrical or magnetic asymmetry of the rotor caused by damage of the bars, while the right sideband $(1+2s)f_r$ occurs due to variations in the rotor speed. Eccentricity of the gap between rotor and stator occurs in the manufacturing process or during operation due to load imbalance or motor and load axis misalignment (horizontal, vertical, radial or combined). These phenomena cause damage to the bearings and motor chassis, which disrupts the symmetry of the gap between rotor and stator. Over time, this leads to increasing the radial force that eventually causes the failure of induction motor. The features for detecting the eccentricity are amplitudes of signal spectrum at sideband frequencies, defined by the expression

$$f_{ecc} = 2f_s \pm f_r \tag{21}$$

where f_s represents frequency of the voltage source, and f_r is frequency of rotation.

By combining the foregoing features of the vibration signal in time and frequency domain, we can form a set of characteristic features that represents the starting point for the detection and diagnosis of faults. The observed squrrel-cage inductionmotor has nominal power of 1.5 kW,nominal speed of 2,860 min^{-1}and nominal current of 5.65 A. This engine was used with two rotors, one of which was correct, and the other one had adamaged rotor bar.

The signals of radial and axial vibration were collected, with two load levels.Data collection wasrepeated for three different positions of the sensor. For each available rotor, load level and sensors position, totally5 signals of axial and radial vibration were recorded. Vibration measurement was conduscted using two uniaxial piezometric accelerometers with high sensitivity (100 mV/g). Accelerometers were attached to the motor using megnetic stands. This mounting methodis very often used for mobile measuring equipment, as it provides fast and simple equipment instalation (Kanović et al., 2012).

Accelerometers were connected with a specialized 24-bit data acquisition card NI-9234 , which provides excitation of accelerometers with2 mA current andvoltage signal standardization inrange ± 5V. The card has four channels and a maximum sampling frequency of 51.2 kHz. Through the USB port it was connected to a portable computer, and LabView package software was usedfor signal processing. Software package *MATLAB*, i.e., itsStatistics and Optimization Toolboxwas used for computational statistical analysis.

In the present case study, the featuresin the frequency domain, i.e., left and right sidebands of the first order were calculated for all collected signals. These features represent inputs in PBRC algorithm.

Figure 3 shows the moment whensymptoms of broken rotor bars occur (red line marked).

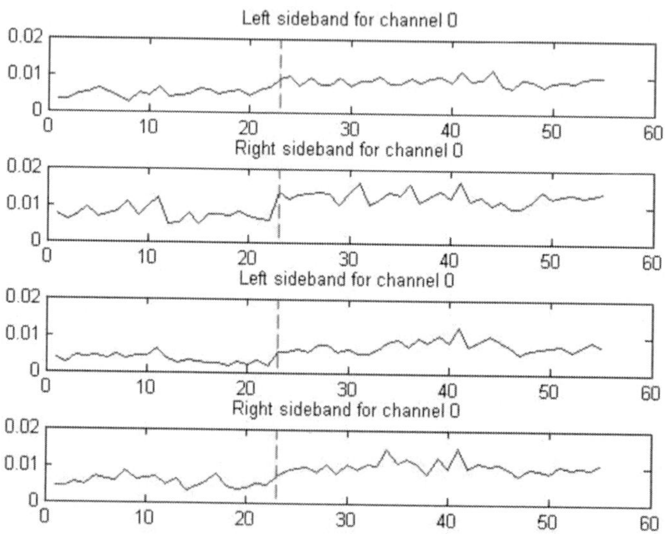

Figure 3. Left and right sidebands of the first order (channel 0 and channel 1), and moment of occurence of fault.

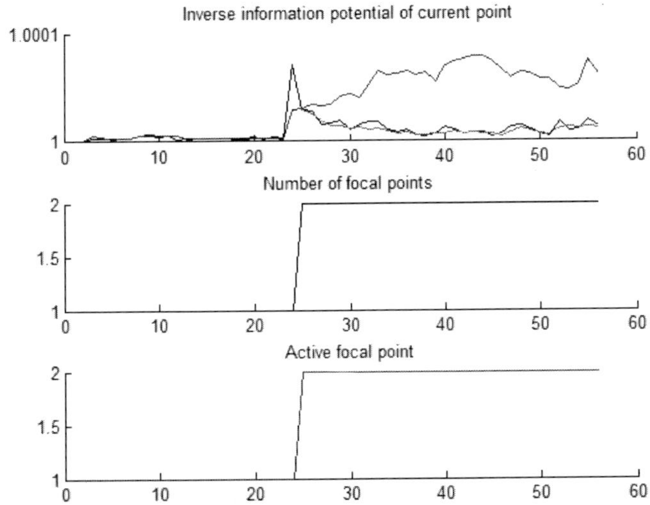

Figure 4. Inverse information potential of the current feature vector, the inverse potentials of all focal points, number and active focal point.

Figure 4 shows the resulting value of the inverse information potential of the current feature vector and the inverse potentials of all focal points. Operating regime with healthy motor was described by the first focal point that is active from the beginning of the measurement until the appearance of fault. The figure clearly shows tha,t when the fault occurs, the algorithm successfully recognizes the change of operating regime and the second focal point becomes active, which describes the operating regime of the motor with damaged rotor bar.

In the second study we observed a part of complex industrial plant in Victoria Oil sunflower oil plant in Sid, Serbia. The whole plant roughly consists of three subsystems: peeling, pressing and extraction subsystem. In our analysis, only the subsystem for extraction was considered. More precisely, our main interest was focused on the desolventizer, which is a part of extraction subsystem. The extracted meal at the outlet of the extractor is conveyed to the desolventizer, where it is heated and stripped with life steam in order to remove the solvent. A toasting effect is also achieved by which, mainly for soybean, anti-nutritional factors in the meal, such as trips in inhibitors, are inactivated. The desolventizing and toasting section is combined with a meal drying and cooling section within the same vessel. The variable of main interest in this process is the temperature of water steam and hexane mixture, which enters the scrubber. That temperature is important because the scrubber is a part of explosive zone and the temperature should never exceed 90°C. Currents and velocities of rotary feeders of the second, third, fourth and fifth floor are taken into consideration, with the rotary feeder velocity (7603.2) of second floor having the most influence on temperature of mixture. Other variables that affect mixture temperature are temperature of meal at the entrance of desolventizer and meal levels on the second, third, fourth, fifth and eighth floor of desolventizer. Further, very important signal is the current of mixer which is used to mix meal in every floor of the desolventizer.

Whenever meal enters desolventizer, it is mixed by mixer (7603), which is unique for all floors. It is of great importance to observe the current of the mixer. Furthermore, from inlet in the part of fifth floor hot water steam is inserted by valve (TV 70DTDC - S). That steam is used to dry out the meal. Due to the fact that steam dries out the meal, its temperature drops and when it mixes with hexane from the meal it exits the desolventizer on its top left side and that mixture then enters the scrubber (7640). Temperature of mixture of hexane and steam should not exceed 90°C, because it would cause an explosion. The problem that occurs periodically is that when not enough meal is inserted in the desolventizer, the hot steam with larger temperature mixes with hexane and then enters the scrubber. Therefore we have to observe the variables which are important of holding the temperature of this mixture of steam and hexane within normal boundaries. The variables that mostly influence to this mixture are the *rotary feeder* (7603.2) *velocity and mixer current.* Rotary feeder (7603.2) velocity provides us indirectly with the data on quantity of meal in the system. Mixer current enables us to realize if meal is mixed in proper way, meaning indirectly, if enough steam is used to dry out the meal or too much of the steam, mixed with hexane is transmitted into the outlet with larger temperature than usual. To deal with this kind of problem, it is necessary to predict in advance as much as possible when this event will occur, and then treat it in some way, either injecting more meal into the desolventizer or stopping it completely.

Designed expert system can be represented with the following scheme:

First essential part of expert system is using PBRC in order to find inverse potentials of two separate signals, mixer current and rotary feeder velocity. Both of these signals at the input of each of these PBRC procedures were filtered with first order filter. PBRC was applied twice, each time for other signal and output was acquired as following. For input signal, mixer current, two output signals were acquired, inverse potential of nominal state (P_I^{-1} (nominal)) and inverse potential of faulty state (P_I^{-1} (faulty)). Based on these two potentials it is possible to predict the moment when the fault is going to occur. This can be done by the fact that the point when inverse potential of faulty state becomes of less value than inverse potential of nominal state is proclaimed as the point when fault is detected. Same approach was applied for the other signal, rotary feeder velocity. As a result the following potentials were acquired. It can be seen on the graphs that the point of intersection of these two potentials gives us a prediction of fault of roughly 400 seconds. It is important to notice that on the graphs were represented both inverse potentials and their derivatives for the reason that they will be used as inputs in fuzzy logic part of expert system. Middle layer of expert system shown in Figure 6consist of three fuzzy logic parts and output layer has one fuzzy logic part. First two parts of middle layer of expert system have the same structure of inputs and outputs, but for different signals. Based on output signals from PBRC we construct fuzzy logic for two separate fuzzy elements in middle layer. Third fuzzy element in middle layer of expert system represents expert knowledge of the process.

Fuzzy Logic of Middle Layer

First two fuzzy logic elements in the middle layer, mixer current fuzzy logic and rotary feeder fuzzy logic are of the same structure. Therefore, only one element will be explained.

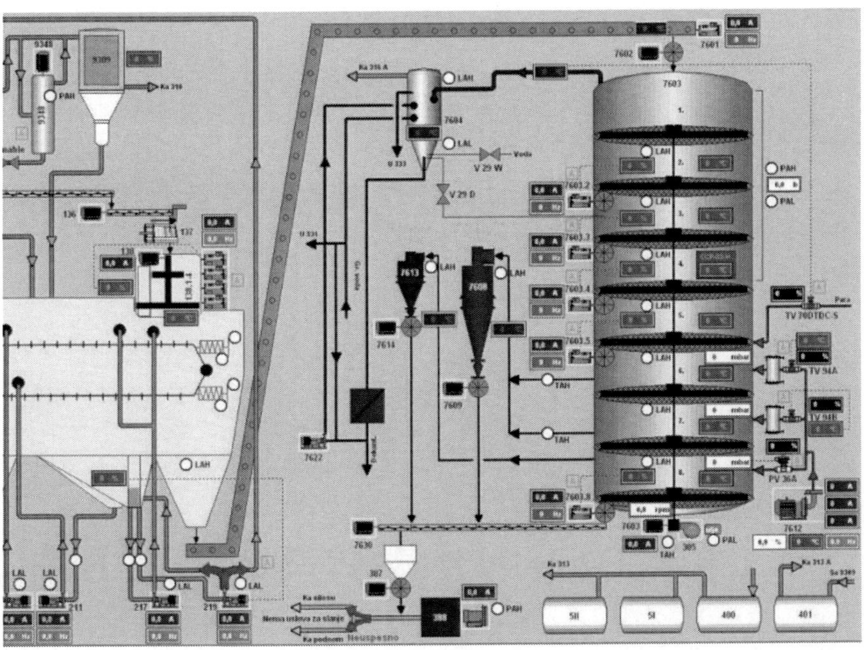

Figure 5. Scada screenshot of the observed subsystem.

PBPC procedure with its inverse potential outputs gives us a very good prediction of some malicious (unwanted) state. Although we have prediction 400 seconds ahead, we tried to interpret these results in order to, hopefully, get prediction even earlier, if possible. Main idea was to use Fuzzy Inference System (FIS) for interpretation of inverse potentials. Hypothesis was that we can see is that inverse potentials are closing up to each other before they intersect. For that we used three signals as FIS inputs. First signal is difference from inverse potential of nominal state signal to inverse potential of faulty state signal (Figure 9 blue line). One can see that prediction of unwanted system state starts when this difference fall below zero. Then we introduce two other input signals as first derivatives of two inverse potential signals. These derivatives are supposed to give us information of either inverse potentials are falling or rising.

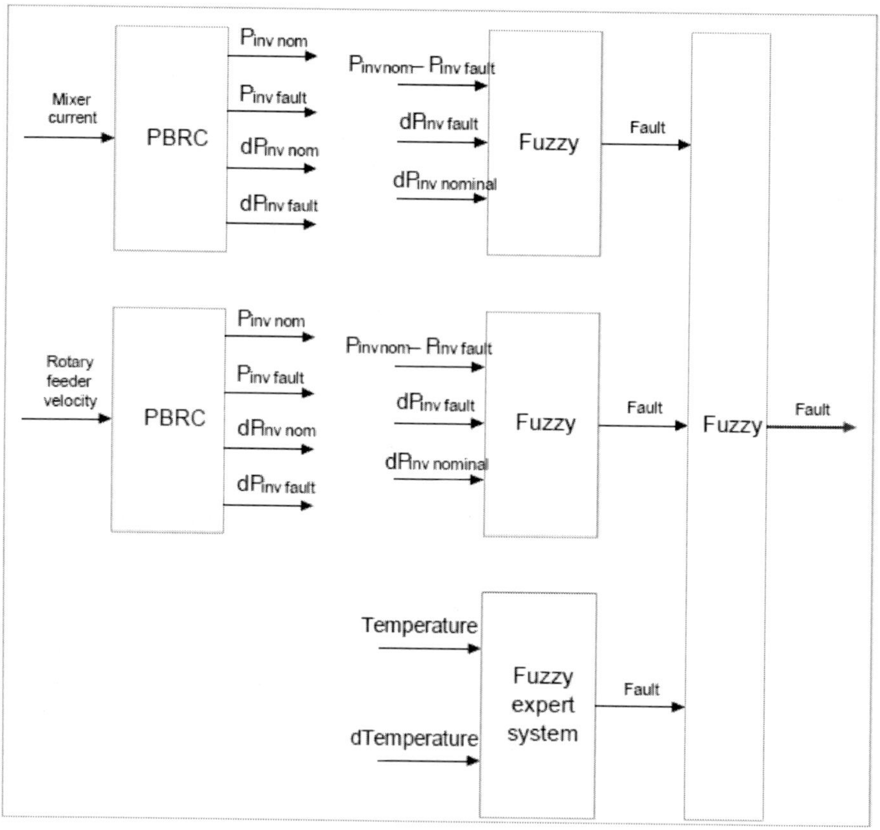

Figure 6. Expert system scheme.

Fuzzy Logic of Output Layer

Output layer of expert system has only one fuzzy logic element, which has three inputs. Every input in this fuzzy element is the output from fuzzy elements in the middle layer. So inputs are fault detection signals based on mixer current, rotary feeder velocity and combination of mixture hexane/steam temperature and its derivative. Output from this fuzzy element is the final fault detection signal.

Finnaly output from every element of this expert system is shown in the following Figure 8.

Fuzzy logic expert system element based on temperature and its derivative has detected Fault 1, and has not detected Fault 2. Time difference between fault and fault detection is negligible.

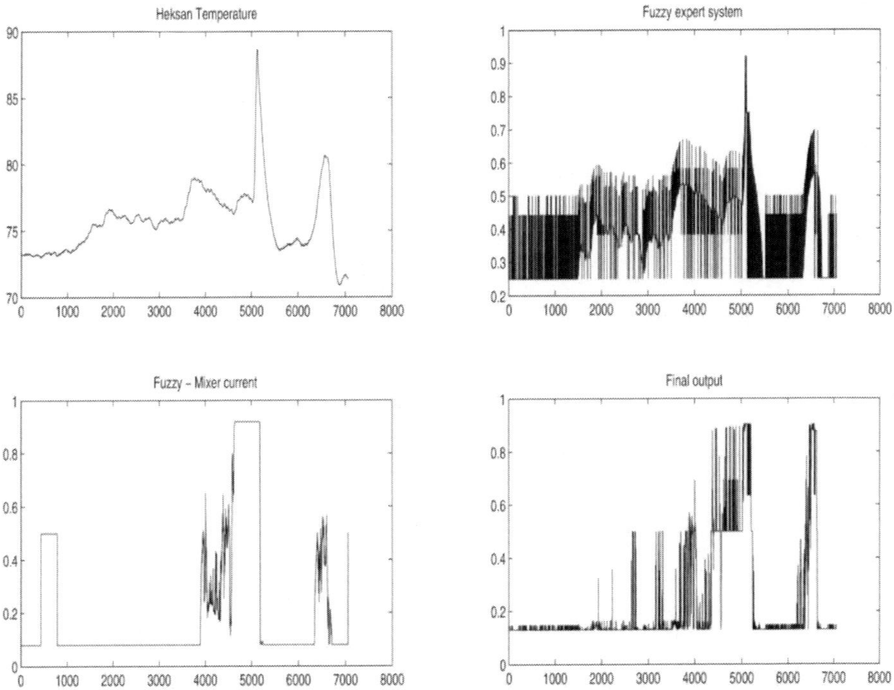

Figure 7. Outputs from every element in the expert system.

Fuzzy logic mixer current element has detected fault 1 about 400 seconds earlier, which is sufficient to react in proper manner, but fault 2 has not been detected.

Fuzzy logic rotary feeder velocity element has detected more than one false fault. Considering only the detection of real fault we conclude that this element has detected fault 1 occurrence 10-20 seconds earlier. In the case such late detection the only action that can be performed is shutting down of whole subsystem. Fault 2 has been detected as early as fault 1, but due to the lower temperature performed action could be different from shutting down the whole subsystem.

Result from the output layer which represent real output of proposed expert system has detected both faults with enough time to react properly without being obligatory to shut down the subsystem. Moreover, it has not detected any false faults.

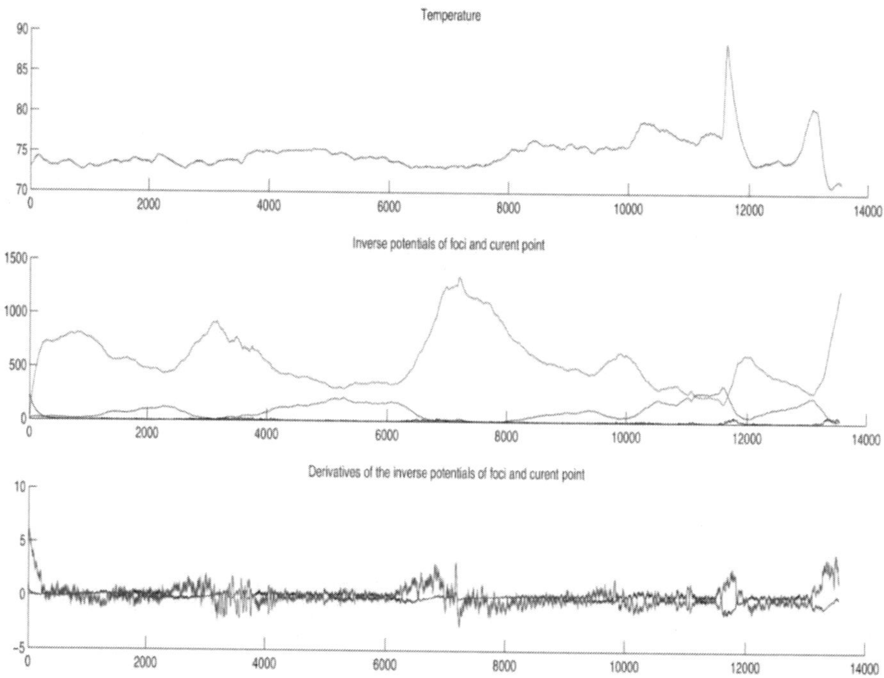

Figure 8. Graph of acquired inverse potentials for mixer current signal along with their derivatives.

Conclusion

In this chapter a self- adaptive expert system for process monitoring and fault detection based on adaptive recursive clustering algorithm (PBRC) is presented.

The key underlying concept of PBRC procedure is the notion of the information potential (3) defined by means of exponentially weighted average distance of the current feature point with respect to the set of all previously seen features. The information potential defined in such way allows for recursive computations and thus, real-time, on-line implementation.

This system is intended to be used as practical diagnostic tool for condition monitoring and fault, malfunction and failure prevention in industrial systems. The system is tested in real industrial environment and obtained results proved its accuracy and efficiency. We showed that using expert system method for failure detection is reliable, fast and robust.

References

Angelov P. P., Filev D. P. (2004). "An Approach to Online Identification of Takagi-Sugeno Fuzzy Models,"IEEE Transaction on Systems, Man, and Cybernetics – Part B: *Cybernetics*, 34(1).

Angelov P.P., Filev D.P., Kasabov N. (2010). "Evolving Inteligent Systems – Theory and Applications," John Willey and Sons, Inc.

Astrom K.J., Wittenmark B.(1997). Computer Controlled Systems – Theory and Design, Prentice Hall.

Becraft W., Lee P.(1993). An integrated neural network/expert system approach for fault diagnosis. *Computers and Chemical Engineering*, 17, 1000-1014.

Ding S.X. (2008). "Model-based fault diagnosis techniques: design schemes, algorithms, and tools," Springer Verlag, Berlin Heidelberg.

Izenman A.J. (2008). Modern Multivariate Statistical Techniques: Regression, Classification and Manifold Learning, Springer Science + Business Media.

Kanović Ž., Matić D., Jeličić Z., Petković Milena. (2013). Induction Motor Fault Diagnosis Based on Vibration Analysis – a Case Study. *Journal on Processing and Energy in Agriculture*, 17(1),47-50.

Kecman V. (2001). "Learning and soft-computing: support vector machines, neural networks and fuzzy logic models," MIT Press.

Kulić F., Kanović Ž., Petković Milena. Matić D. (2011). Expert System for Induction Motor Fault Detection. *Journal on Processing and Energy in Agriculture*, 14(4) 173-177.

Matić D., Kulić F., Bugarski V. (2010). Survey of the methods for online broken bar induction motor fault detection. *Journal on Processing and Energy in Agriculture*, 14(2) 90-92.

Petković M., Rapaić M. R., Jeličić Z.D., Pisano A. (2012) On-line adaptive clustering for process monitoring and fault detection. *Expert Systems with Applications*, 39, 10226-10235.

Rapaic RM., Z.D. Jelicic, A. Pisano, E. Usai. (2010) Second-order sliding modes and soft computing techniques for fault detection, Proc. of the 8th European Workshop on Advanced Control and Diagnosis *ACD 2010*, Ferrara, Italy, pp. 271-277.

Tarifa E., Scenna N. (1997). Fault diagnosis, direct graphs, and fuzzy logic. *Computers and Chemical Engineering*, 21, 649-654.

Venkatasubramanian V. (2003a). "A review of process fault detecton and diagnosis, Part I: Quantitative model-based methods," *Computers and Chemical Engineering*, 27:293-311

Venkatasubramanian V. (2003b). "A review of process fault detecton and diagnosis, Part II: Qualitative models and search strategies," *Computers and Chemical Engineering*, 27:313-326.

Venkatasubramanian V. (2003c). "A review of process fault detecton and diagnosis, Part III: Process history based methods," *Computers and Chemical Engineering*, 27:327-346.

Biographical Sketch

Milena Petkovic, PhD

Affiliation: Department of Computing and Control, Engineering, Faculty of Technical Sciences, University of Novi Sad

Education: PhD in Electrical Engineering and Computer Sciences

Business Address: Trg Dositeja Obradovica 6, 21000 Novi Sad, Serbia

In: Fault Detection
Editor: Daniel Martin

ISBN: 978-1-53610-345-8
© 2017 Nova Science Publishers, Inc.

Chapter 5

AN INTEGRATED FDD SYSTEM FOR HVAC&R BASED ON VIRTUAL SENSORS

Woohyun Kim*[*]*, PhD
Pacific Northwest National Laboratory, Richland, WA, US

Abstract

According to the US Department of Energy, space heating, ventilation and air conditioning system account for 40% of residential primary energy use and for 30% of primary energy use in commercial buildings. A study released by the Energy Information Administration indicated that packaged air conditioners are widely used in 46% of all commercial buildings in the US This study indicates that the annual cooling energy consumption related to the packaged air conditioner is about 160 trillion Btus. Therefore, an automated FDD (Fault Detection and Diagnostics) system that can automatically detect and diagnose faults and evaluate fault impacts has the potential for improving energy efficiency along with reducing service costs and comfort complaints. The primary bottlenecks to diagnostic implementation in the field are the high initial costs of additional sensors. To prevent those limitations, virtual sensors with low cost measurements and simple models are developed to estimate quantities that would be expensive and or difficult to measure directly. The use of virtual sensors can reduce costs compared to the use of real sensors and provide additional information for economic assessment. The virtual sensor can be embedded in a permanently installed control or monitoring system and continuous monitoring potentially leads to early detection of faults. The virtual

[*] E-mail address: Woohyun.kim@pnnl.gov; Address: Pacific Northwest National Laboratory, Richland, WA 99352 US (Corresponding author).

sensors of individual equipment components can be integrated to estimate overall diagnostic information using the output of each virtual sensor.

Keywords: FDD, HVAC&R, fault impact, virtual sensor

Introduction

Based on a survey and analysis of 503 rooftop air conditioners (RTUs) conducted by Cowan (2004), 54% of RTUs were found to have problems including 42% improper airflow, 72% improper refrigerant charge and 20% failed sensors. Another field study released by NBI (2003) based on a total of 215 Heating, Ventilation, and Air conditioning (HVAC) units at 75 sites indicated 46% improper refrigerant charge and 39% low air flow. These problems impact building electrical energy performance by an estimated 8%. A study conducted by Messenger (2008) indicates that unitary air conditioners typically do not achieve rated efficiency because of improper installation or lack of servicing in the field. This paper suggested that service and replacement programs could yield energy savings on the order of 30 to 50%. A study from ADM (2009) evaluated 109 residential units in the field and found that 89 had fault conditions, with 31 having two or more faults. The average energy efficiency ratio for the units increased from 6.6 before servicing to 7.0 after servicing, an average increase of 6.1%.

Braun (2003) presents the automated Fault Detection and Diagnostics (FDD) systems in the field of HVAC and refrigeration (HVAC&R) that have the potential to reduce operating costs by lowering service and energy utility costs. A business productivity is also improved based on the reduction of equipment downtime. In order to be cost effective, an automated FDD system for HVAC&R applications should have low installation cost and low-cost reliable sensors. In order to accomplish this goal, automated FDD systems could be integrated into individual equipment controllers, and provide on-line monitoring, fault identification, and the diagnostic outputs with sufficient information to choose an appropriate action. The proper maintenance based on automated FDD systems can result in significant cost and energy savings that would have an economic and environmental impact.

Katipamula and Brambley (2005a and 2005b) conducted a detailed review of automated FDD studies of building systems and published a two-part review that summarized the automated FDD and prognostics method. Their intent was to increase research and development community awareness of the body of automated FDD and prognostics developments in other fields as well as advancements in HVAC&R applications. The first part of the review focused on generic automated FDD and prognostics, provided a framework for categorizing methods, described them, and identified their primary strengths and weaknesses. The second part of the review focused on research and applications specific to the fields of HVAC&R, briefly discussed the current state of diagnostics in buildings, and discussed the future of automated diagnostics in buildings. Most of the earlier work involving FDD systems has involved the use of residuals between measurements and expected values from models for state variables.Once a fault is detected, then a fault diagnostic classifier is employed to find the best match between the changes in residuals and a set of rules associated with different types of faults.One difficulty in applying this approach with a fault diagnosis classifier is in handling multiple faults that occur simultaneously because the state variables can depend on more than one fault along with the operating conditions. The FDD system should be able to decouple these effects in order to handle multiple simultaneous faults.

Li and Braun (2007a) presented a decoupling FDD approach that relies on the use of virtual sensors that are uniquely dependent on individual faults and decoupled from the effects of other faults.Virtual sensors are used in place of real sensors in most cases because the measurements required for the decoupling features would be prohibitively expensive (e.g., refrigerant mass flow rate). The FDD system based on virtual sensors can be significantly more sensitive to system malfunctions and easier to implement than the previous FDD methods. Once faults are detected and the causes of the faults are identified, proper action should follow to fix the problems, adapt the control, or flag them for continued monitoring. The fault impact is used to determine thresholds for the FDD system for decision support. An assessment of the severity of the fault isessential to this decision process and virtual sensors can be used as inputs to this analysis. In selecting the optimal thresholds based on no-fault data, there is a tradeoff between sensitivity and false alarms. The fault diagnostic methods based on virtual sensors are combined with fault detection and no separate diagnostic classification is necessary. This research describes a complete implementation and demonstration of the decoupling FDD system applied to RTU that incorporates integrated virtual sensors and fault impact model.

Review of Virtual Sensors

A virtual sensor uses low-cost measurements and a simple mathematical model to estimate a quantity that would be expensive and/or difficult to measure directly.

The use of virtual sensors can reduce costs compared to the use of real sensors and provide additional information for economic assessment. Virtual sensors are now being widely used in various fields including automobile, wireless communication, and sensor networks, (Lichuan Liu et al., 2009) but have seen limited application for HVAC&R equipment.

The development of virtual sensors for the HVAC&R field has been slow due to limitations, such as the fragmented nature of the industry and the emphasis on initial costs (Yu, Li, and Braun, 2011). Recently, Li and Braun (2007a, 2007b, 2009a, and 2009b) developed a number of virtual sensors for packaged air conditioning equipment. The virtual sensors were developed for the compressor, expansion valve, condenser, evaporator, and refrigerant charge.

Benefits of Virtual Sensors

There are several benefits of virtual sensors over real sensors. First, some quantities cannot be measured directly due to slow feedback time. Some real sensors may not respond as fast as what is needed for process control or feedback to the operators. Virtual sensors can be used to resolve these issues by predicting sensor outputs using mathematical models to provide continuous information from time-delayed input signals. There can also be significant lags due to communication between readings (Wilson, 1997). This lag time can prohibit a proper control and operator response. Virtual sensors can predict the sensor outputs without any time-delayed outputs. Virtual sensors are also useful when the installation of a physical sensor is not possible due to too small a space. Moreover, geometry limitations can often cause inaccuracy of real sensors.

Second, a popular use for virtual sensors is in the area of fault tolerance. If a real sensor is damaged during a process, a virtual sensor can be used until the real sensor can be brought back online. The output of the virtual sensor also can be compared to the output of the real sensor in order to detect a fault. With the use of virtual sensors, a control system can detect a fault, localize the fault within the system, and limit the propagation of the fault until the system is able to recover. For example, a virtual sensor was used to provide data redundancy in a helicopter aircraft application (Heredia & Ollero, 2010). The virtual sensor was able to predict unfolding wings when a helicopter aircraft was taking off. As the wings unfold, the controller can transfer motion from the rotor to a propeller to proceed in foreword flight as an airplane.

Third, virtual sensors can significantly reduce costs by eliminating the need for some expensive physical sensors. Some sensors are extremely expensive and physical sensors require maintenance throughout their lifespan. The physical sensors can be replaced with a combination of models requiring lower cost sensors as inputs. Some sensors also have issues maintaining their calibration due to their design or operating environment. In these cases, a virtual sensor model can be trained using data from a freshly calibrated sensor and used in place of it.

Lastly, virtual sensors can encapsulate real sensor information and provide for simpler interfaces. A virtual sensor can increase usability of the sensor data and decrease the overhead of learning different interfaces. For example, a mobile virtual sensor, a processing middleware, was designed for tracking objects through several cameras in a smart application (Kumar et al., 2008). Through the use of mobile virtual sensors, CPU load was reduced by 60% because the sensors allowed for more selective processing of the image data.

General Steps for Developing Virtual Sensors

The general process of developing a virtual sensor can be defined using three main steps, shown in Figure 1. Data collection and pre-processing are the fundamental steps for accurate and reliable virtual sensor models. The data collection provides calibrated measurements as input data to pre-processing. The pre-processing algorithms select the type and range of input data based on the modeling approach. A virtual sensor model that is based on steady-state operating conditions should filter out transient data from the measurements using a steady state detection algorithm (Li & Braun 2007a; Wichman & Braun 2009).

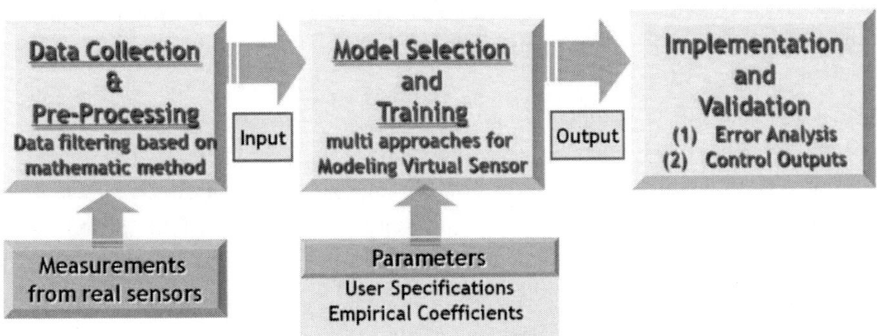

Figure 1. General steps for developing virtual sensors.

Second, model selection and training is the core process in developing virtual sensors. Based on the selected model, physical parameters and input data requirements can be specified. In most cases, the model may be trained based on measured input data and real sensor outputs. Finally, the virtual sensor can be implemented as a standalone sensor or embedded within a system. A standalone sensor has its own hardware, embedded software, and input/output channels. The implementation of a virtual sensor can then be validated based on statistical error analysis.

Overview of Virtual Sensor Developments in Other Fields

A literature review of virtual sensors developed in other fields is useful in identifying the potential opportunities and research needs for virtual sensing technology in the HVAC&R field. Virtual sensing techniques could facilitate the development of more cost-effective and robust diagnostic systems and optimal control. In particular, virtual sensing has found widespread applications in process controls and automobiles, so this section focuses on these two fields in providing a brief history of notable developments.

Virtual Sensing in Automobiles

A large number of virtual sensors have been developed during the past several decades for automobile applications. The automobile is equipped with numerous sensors to detect the operating status and increase the robustness of the automobile (J. Stéphant et al., 2004). In industry, the development of virtual sensors has enabled cost effective solutions for optimizing driving performance, safety, functionality, and reliability of vehicles (Healy, 2010). For example, virtual sensors were used to estimate the tire load friction of the automobiles. The longitudinal wheel slip and the wheel torque were measured from standard sensors. A virtual sensor using a Kalman filter estimates a linear relation between wheel slip and torque, and the slope of the line is mapped to a surface in a classifier (Gustafsson, 1998). In addition, virtual velocity sensors have been developed that are used to provide travel time and speed measurements of arterials and freeways based on probe sensor locations (Daily, 2002).Figure 2 shows examples of virtual sensors for a "systemized" vehicle as described by Yu et al. (2011).

Virtual Sensing in Control Systems

Virtual sensors have been widely used in process control engineering software since the early 1980s. Typical virtual sensors in process control engineering (Rallo etal., 2002, Kabadayi etal., 2006) use dynamic models and are often applied as an observer for feedback control. Virtual sensors have also been studied actively by researchers for smart robot systems. New system architectures have allowed for the collaboration of both real and virtual robots, which utilize both real and virtual sensor data (Dixon et al., 1999). Virtual sensors provide sensing for virtual robots as well as to augment existing onboard sensors on a real robot with virtual data. Many tools were also created to allow multiple users access to visualize robot information and control their systems.

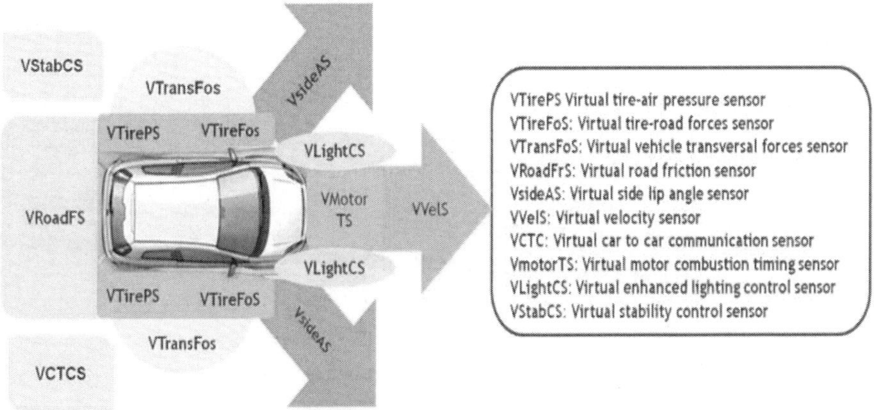

Figure 2. Virtual sensors for systemized vehicle (from Yu, et al., 2011).

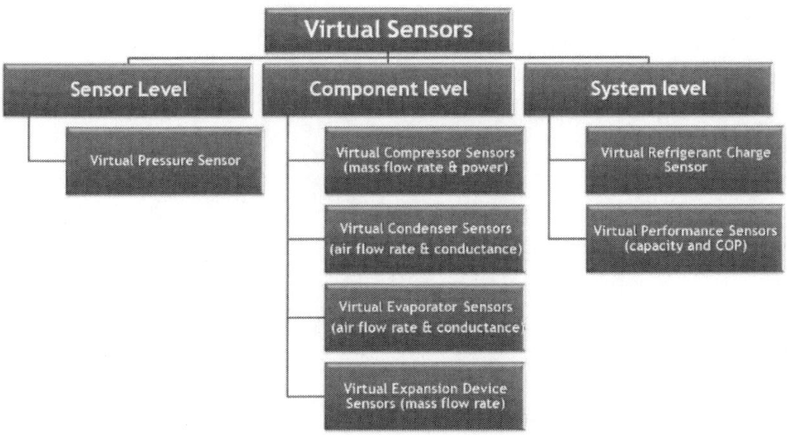

Figure 3. Virtual sensor classifications for HVAC&R (Li and Braun, 2009a).

Development of FDD System Based on Virtual Sensor for HVAC&R

The primary objective of this chapter is to present a development of virtual sensors for HVAC&R system. The virtual sensors can be divided into three classes, sensor-level, component-level, and system-level as depicted in Figure 3 (Li and Braun, 2009a). For instance, a vritual component-level sensor can be used to estimate refrigerant mass flow rate and power consumption of compressor using only temperature sensors (along with virtual sensor level sensor). These quantities can be used to determine system capacity and energy efficiency as part of fault impact evaluation. They can also be used in combination with other virtual sensors to isolate specific faults. Various virtual sensors for HVAC&R are reviewed in this section.

Sensor Level Virtual Sensor

The sensor level provides virtual sensors that replace real measurementsusing lower cost measurements and correlations that do not depend on component performance (Li and Braun, 2009c). For example, virtual pressure sensors estimate condensing refrigerant pressure (P_c) and evaporating refirgerant pressure (P_e) using refrigerant condensing and evaporating saturation temperature measurement and refrigerant property relations.

Component Level Virtual Sensor

The component level virtual sensors utilize component models and low cost measurements to determine quantities that can be used for fault detection. Faults can be detected by comparing fault-free expected values to current values based on measurements, and analyzing their residuals. Virtual sensors provide estimates of current features that can be compared to expected values for fault-free operation.

Virtual Refrigerant Mass Flow Sensor

Refrigerant mass flow rate is an important measurement for monitoring equipment performance and enabling automated FDD system. However, a traditional mass flow meter is expensive to purchase and install. Three different Virtual

Refrigerant Mass Flow (VRMF) sensors are presented and evaluated that use mathematical models to estimate flow rate using low cost measurements (Kim and Braun, 2016). Outputs from the three VRMF sensors can be compared in order to detect a fault and localize faults within a system, including loss of compressor performance and faulty expansion device. The empirical coefficients for the VRMF sensor were determined from normal data was applied to predict refrigerant mass flow rates for all of the available data including various fault conditions.

The first model uses a compressor map that relates refrigerant flow rate to measurements of condensing and evaporating saturation temperature, and inlet temperature measurements as shown in equation 1 (ANSI/ARI, 2004).

$$\dot{m}_{predicted1} = \rho_{suc} \cdot (a_0 + a_1 \cdot T_c + a_2 \cdot T_c + a_3 \cdot T_e^2 + a_4 \cdot T_e^2 + a_5 \cdot T_c \cdot T_c + a_6 \cdot T_e^3 + a_7 \cdot T_e^3 + a_8 \cdot T_e^2 \cdot T_c + a_9 \cdot T_c^2 \cdot T_e) \quad (1)$$

where $\dot{m}_{map}$ is the estimated refrigerant mass flow rate, the a's are empirical coefficients, T_e is the evaporating saturation temperature, T_c is the condensing saturation temperature, andρsuc is the density at the inlet of the compressor. The coefficients of model are estimated using a regression analysis applied to experimental data.

The second model uses a compressor energy balance with the power consumption from a virtual sensor, $\dot{W}_{predicted}$, and compressor heat loss model, α_{loss}, as shown in equation 2 and 3. The power consumption from a virtual sensor is used to estimate compressor input power using the inlet and outlet saturation temperatures as shown in equation 4.

$$\dot{m}_{predicted2} = \frac{\dot{W}_{predicted} \cdot (1 - \alpha_{loss})}{(h_{dis} - h_{suc})} \quad (2)$$

$$\alpha_{loss} = b_0 + b_1 \cdot P_c + b_2 \cdot P_e + b_3 \cdot T_c + b_4 \cdot T_e \quad (3)$$

$$\dot{W}_{predicted} = \rho_{suc} \cdot (c_0 + c_1 \cdot T_c + c_2 \cdot T_c + c_3 \cdot T_e^2 + c_4 \cdot T_e^2 + c_5 \cdot T_c \cdot T_c + c_6 \cdot T_e^3 + c_7 \cdot T_e^3 + c_8 \cdot T_e^2 \cdot T_c + c_9 \cdot T_c^2 \cdot T_e) \quad (4)$$

where the b's and c's are empirical coefficients, h_{dis} and h_{suc} are the discharge line and suction line refrigerant enthalpies.

The third model is developed using an empirical correlation for thermal expansion valves (TXV) based on an orifice equation, as show in equation 5 and 6. TXV as expansion device are used to drop the pressure of the refrigerant and to regulate the refrigerant mass flow rate in response to a variable load.

$$\dot{m}_{predicted3} = \left(d_1 \cdot (P_e - P_{suc})^2 + d_2 \cdot (P_e - P_{suc}) + d_3\right) \cdot \sqrt{\left(2 \cdot \rho_f \cdot [P_c - P_e \cdot K]\right)} \quad (5)$$

$$K = e_1 \cdot \left(\frac{T_{SC} + 2}{T_{cri}}\right)^{e2} + e_3 \cdot \left(\frac{P_{cri} - P_e}{P_{cri}}\right) + e_4 \quad (6)$$

where the d's and e's are empirical coefficients, ρ_f is the density of the refrigerant at the valve inlet, T_{SC} is the subcooling of the refrigerant at the valve inlet, P_{suc} is the suction pressure of compressor, T_{cri} and P_{cri} are the critical temperature and pressure and $D_{orifice}$ is the orifice diameter. The empirical coefficients of model are determined based on the available normal test data using regression.

Virtual Air Flow Rate Sensor

The condenser heat exchanger without service for several years may become dirty, inhibiting heat transfer from the refrigerant side to the air side. Sometimes, the outdoor air flow may also decrease because of a defective fan motor, or a poorly installed controller. However, air flow measurements are generally very expensive and unreliable for application in the field. The Virtual Air Flow Rate (VAF) sensor for condenser can be compared with the target flow to detect condenser faulty condition (Kim, 2016). The target flow can be obtained from a manufacturer's catalog or from a normal value when the FDD scheme is implemented assuming that there is no fouling. The air flow rate can be estimated using energy balances on the condenser as expressed in equation 7.

$$\dot{V}_{predicted} = \frac{\dot{m}_{predicted}\left(h_{dis} - h_{liquid,in}\right)}{\left(T_{a,oc} - T_{a,ic}\right)} \frac{v_{cond,a}}{C_{P,air,cond}} \quad (7)$$

where $V_{predicted}$ is the condenser air volume flow rate, $v_{cond,a}$is the condenser air specific volume, $C_{p,air,cond}$ is the air specific heat, $T_{a,oc}$ is the condenser outlet air temperature, $T_{a,ic}$is the condenser inlet air temperature, $\dot{m}_{predicted}$is the refrigerant mass flow rate provided from the VRMF sensor, h_{dis} is the discharge line refrigerant enthalpy, and $h_{li,in}$ is the liquid-line refrigerant enthalpy.

System Level Virtual Sensor

System level virtual sensors provide outputs for quantities that could not be determined solely using component level information, including overall refrigerant charge, capacity, and energy efficiency.

Virtual Refrigerant Charge Sensor

Improper refrigerant charge causes problems in the field, such as compressor damage and a significant energy consumption increase. The Virtual Refrigernat Charge (VRC) sensor (Li and Braun, 2009a) uses a correlation in terms of superheat and subcooling that are determined using four low-cost surface mounted temperature sensors. The charge deviation relative to the rated charge is expressed as

$$\frac{\left(m_{total} - m_{total,rated}\right)}{m_{total,rated}} = \frac{1}{K_{ch}}\left\{\left(T_{sc} - T_{sc,rated}\right) - K_{sh/sc}\left(T_{sh} - T_{sh,rated}\right)\right\} \qquad (8)$$

where m_{total} is the actual total charge, m_{rated} is the nominal total refrigerant charge, $K_{sh/sc}$ and K_{ch} are two constants that are characteristics of a given system, and $T_{sc,rated}$ and $T_{sh,rated}$ are liquid line subcooling and suction line superheat at rated conditions with the nominal charge. Parameters of the method can be tuned for each specific system based on measurements obtained at different refrigerant charge levels. Based on previous research (Kim and Braun, 2012a), the VRC sensor was found to work well in estimating the refrigerant charge.

Virtual Performance Sensor

Capacity and energy efficiency are important information for evaluating the fault impact for HVAC&R system. Capacity and energy efficiency are calculated based on virtual sensors for mass flow rate and power consumption with some other relatively low cost temperature measurements, shown in equations 9 and 10. An assessment of the severity of the fault is essential to this decision process and virtual performance sensor can be used as inputs to this analysis. This type of information can be used within an online tool for assessing the economics associated with servicing a unit if faults exist.

$$Q_{predicted} = \dot{m}_{virtualsensor} \cdot \left(h_{evap,out}\left(P_{evap,sat}, T_{evap,out}\right) - h_{liquid,out}\left(P_{cond,sat}, T_{liquid,out}\right)\right) \quad (9)$$

$$COP_{predicted} = \frac{Q_{predicted}}{W_{virtualsensor}} \quad (10)$$

where $Q_{predicted}$, and $COP_{predicted}$ are the predicted capacity and energy efficiency using the VRMF ($\dot{m}_{virtualsensor}$) sensor and VCP ($W_{virtualsensor}$) sensor as inputs.

Integration of Virutal Sensors

Widespread application of virtual sensors in HVAC&R would require an integrated, on-line performance monitoring and diagnostic system. To achieve an integrated system, the virtual sensors should have linkages with shared outputs and inputs, and provide real-time information about capacity, power consumption, and energy efficiency for assessing economic impact. Figure 4 shows an example of the inter-relationships between real and virtual sensors within an integrated FDD system for HVAC&R system. Sensor-level virtual pressure sensors estimate condensing and evaporating pressures using saturation temperature measurements and property relations. A component level virtual sensor for refrigerant flow rate uses the outputs of the virtual pressure sensors and a component model. The output of the derived virtual refrigerant flow sensor is used for input to a system level sensor (virtual performance sensor) to calculate energy efficiecny of HVAC&R system.

Fault Impact Model

Fault impact models for capacity ($Q_{expected,map}$)and energy efficiency ($COP_{expected}$) were developed to estimate reference information at different operating conditions. The fault impact model can be used to quantify deviation from the current estimation based on virtual sensors as well as determine whether the current fault is severe enough to warrant service. The fault impact models were correlated in terms of ambient temperature ($T_{air,cond,in}$), indoor dry bulb temperature ($T_{air, evap,dry,in}$) and indoor wet bulb temperature ($T_{air, evap,web,in}$), as shwon in euqaiton 11,

12 and 13. The empirical coefficients of model performance models were estimated based on performance data from available normal existing test data or test data provided by system specification using linear regression.

The outputs of the fault impact modelcan be used as the rated performance and compared with the current calculated performance based on virtual sensor to determine whether service is needed due to fault existence. Therefore, it is very important to define reasonable thresholds of energy impact evaluation for appropriate fault diagnosis. In this research, the threshold for performance degradation was set to 10% based on the fault impact evaluation in (Kim and Braun, 2012b). If performance degradation was less than 10%, the fault impact for performance and cost was considered to be small compared with service costs.

$$\begin{aligned} Q_{expected,map} &= a_0 + a_1 \cdot T_{air,cond,in} + a_2 \cdot T^2_{ari,cond,in} + a_3 \cdot T_{air,evap,web,in} + a_4 \cdot T^2_{ari,evap,web,in} \\ &+ a_5 \cdot T_{air,evap,dry,in} + a_6 \cdot T^2_{air,evap,dry,in} + a_7 \cdot T_{air,cond,in} \cdot T_{air,evap,web,in} \\ &+ a_8 \cdot T_{air,cond,in} \cdot T_{air,evap,dry,in} + a_9 \cdot T_{air,evap,dry,in} \cdot T_{air,evap,web,in} \end{aligned} \tag{11}$$

$$\begin{aligned} W_{expected,map} &= b_0 + b_1 \cdot T_{air,cond,in} + b_2 \cdot T^2_{ari,cond,in} + b_3 \cdot T_{air,evap,web,in} + b_4 \cdot T^2_{ari,evap,web,in} \\ &+ b_5 \cdot T_{air,evap,dry,in} + b_6 \cdot T^2_{air,evap,dry,in} + b_7 \cdot T_{air,cond,in} \cdot T_{air,evap,web,in} \\ &+ b_8 \cdot T_{air,cond,in} \cdot T_{air,evap,dry,in} + b_9 \cdot T_{air,evap,dry,in} \cdot T_{air,evap,web,in} \end{aligned} \tag{12}$$

$$COP_{expected} = \frac{Q_{expected,map}}{W_{expected,map}} \tag{13}$$

Validation of FDD System Based on Virutal Sensor for HVAC&R

System Description and Test Conditions

RTU field test data was used to evaluate the performance of virtual sensors. The field test dataset that included transients associated with on/off cycling of the equipment and the effects of solar, wind, and rain collected by the manufacturer were used to evaluate the performance of virtual sensors. System descriptions are given in Table 1 and testing conditions are listed in Table 2. Table 1 gives specifications for a a 10 ton RTU system where data were obtained through field testing. The systems employed a TXV as the expansion device and R-410A as the refrigerant.

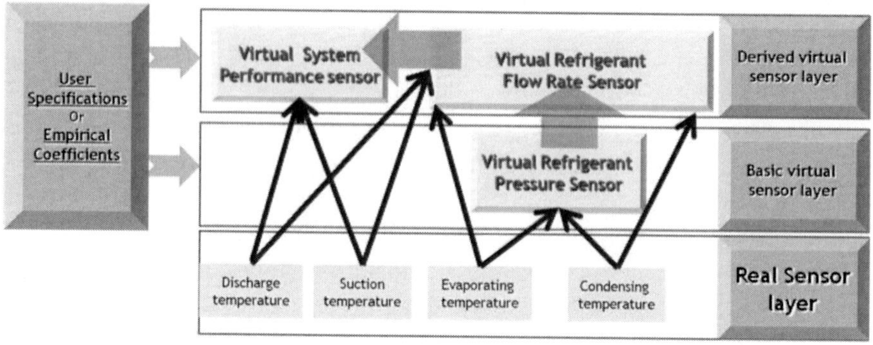

Figure 4. Example of virtual sensor interactions for HVAC&R.

Table 1. Specification of RTU system

Nominal capacity	Refrigerant	Expansion type	Indoor airflow	Outdoor airflow	Total power	Indoor fan	Outdoor fan
[tons]	-	-	[CFM]	[CFM]	[W]	[W]	[W]
7.5	R410A	TXV	2885	5750	6800	1350	497

Table 2. Testing conditions

Indoor air dry temperature	Indoor air wet temperature	Outdoor dry temperature	Condenser block ratio	Refrigerant charge level
[°F]	[°F]	[°F]	[%]	[%]
68 ~ 86	38 ~ 75	61 ~ 97	0, 30, 43, 55	70,75,80, 85,100,140

The ranges of test conditions are given in Table 2. The field tests for RTU system were performed under different refrigerant charge and condenser fouling conditions. Condenser fouling was simulated by blocking a portion of the condenser heat exchange area. The condenser fouling fault level is defined as the percentage of outdoor heat exchanger area blocked by paper strips. Fault levels were considered from 0 to 55%. The refrigerant charge fault level is the charged amount as a percentage of the rated charge. The refrigerant charge was varied from 70 to 140% of normal charge. During the field test period, the outdoor-air temperature was as high as 97°F during the daytime and dropped as low as 61°F. Data for relatively low outdoor temperatures were used to validate the algorithms for conditions that would occur during the off season, when regular maintenance procedures are often performed. The indoor dry/wet bulb temperatures were as high as 86/75°F and dropped as low as 68/38°F.

Virtual Sensor Peformance Evaluation

The accuracy of the virtual sensors(VRC, VRMF, and VAF) sensors was evaluated for RTU test data in terms of the RMS deviation from the actual measurements presented on a percentage basis. Overall, the virtual sensors showed good performance within 10% of real measurements regardless of the different operating temperatures and faulty test conditions. The outputs of virtual sensors can provide information regarding the presence of certain faults and are relatively insensitive to the existence of other system faults.

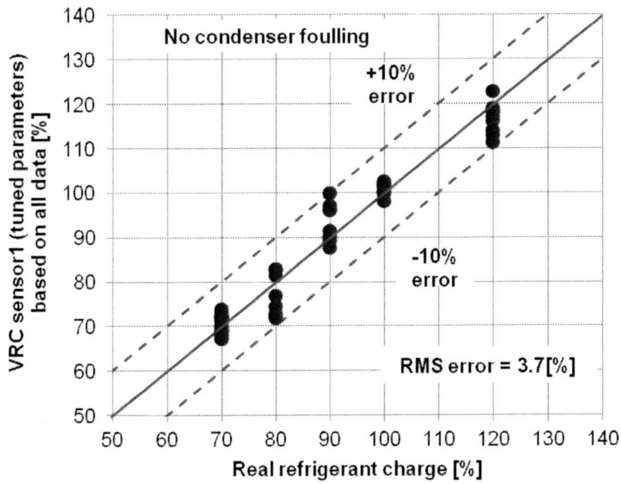

Figure 5. Performance of VRC sensor for RTU II under no condenser fouling.

Validation of Virtual Refrigerant Charge Sensor

VRC sensor was evaluated in terms of RMS deviation between predicted and actual charge levels relative to nominal charges. Also, the existing charging method specified by the manufacturer was applied to RTU system and compared with the VRC sensor. The existing charging approach requires compressor discharge pressure and liquid line temperature measurements to verify refrigerant charge amount. The technician evaluates whether to add or remove refrigerant based on a difference between the measurement and a target value based on the charging chart supplied by the manufacturer. For each charge level, simulated condenser fouling were introduced to the system to test its robustness.

Figure 5 shows the accuracy of the VRC sensor under conditions without heat exchanger fouling. The VRC sensor showed the RMS errors of 4%. The VRC sensors work very well under all refrigerant charge conditions. A relatively accurate prediction of refrigerant charge amount was made even when refrigerant was overcharged at low ambient temperature conditions. Figure 6 shows the results of the existing field charging method for the RTU system without heat exchanger blocking. The solid line indicates the target pressure and temperature combination. The points that are above the line indicate that more refrigerant needs to be added to the system and those below the line indicate that refrigerant needs to be removed. It showed 70~80% as undercharged, 90~100% as nominal, and 120% as overcharged. Overall, the manufacturers' charging method and VRC sensor can provide accurate estimates when no heat exchanger blocking is present.

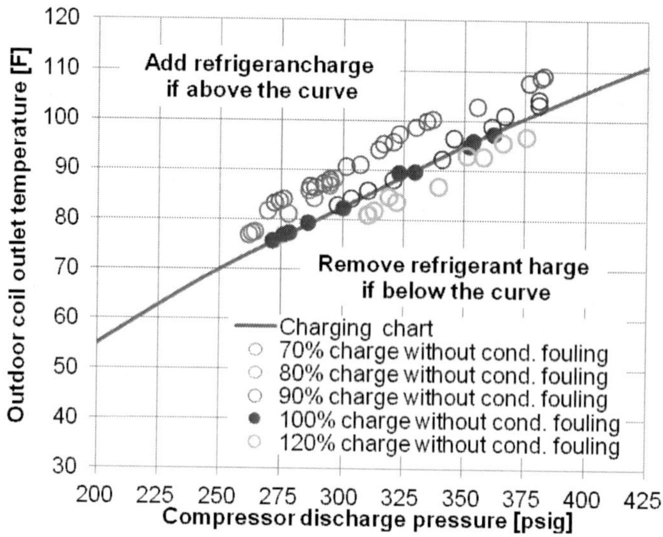

Figure 6. Charging results based on manufacturers' charging method for RTU II under no heat exchanger blocking.

Figure 7 shows the results of VRC sensor for RTU system II with heat exchanger blocking. VRC sensor shows good performance for predicting refrigerant charge levels when condenser fouling conditions exist. Some errors between 80 and 90% refrigerant charge levels were slightly over 10%. These points are associated with large condenser blocking of approximately 50%. The cases where the VRC sensor had difficulty were when the system operated with zero subcooling at over 50% reduction of condenser air flow rate. In this case, the VAF sensor for the condenser would be helpful at predicting the condenser fouling fault. Overall, the VRC sensor provides accurate refrigerant charge estimates regardless of the condenser fouling. Figure 8 shows the charging results when heat exchanger blockage is present. While it showed 70% as undercharged, 80~90% was detected as nominal charge and 100~120% was detected as overcharged. This suggests that when there is condenser fouling the charging method can indicate normal charge even when the unit is undercharged by as much 20%. In addition, the manufacturers' charge verification utilizes pressure gauges or transducers installed at the service valve. The installation of these gauges or transducers can lead to refrigerant leakage. Because of these limitations, the current protocols for checking refrigerant charge may be doing more harm than good in many situations.

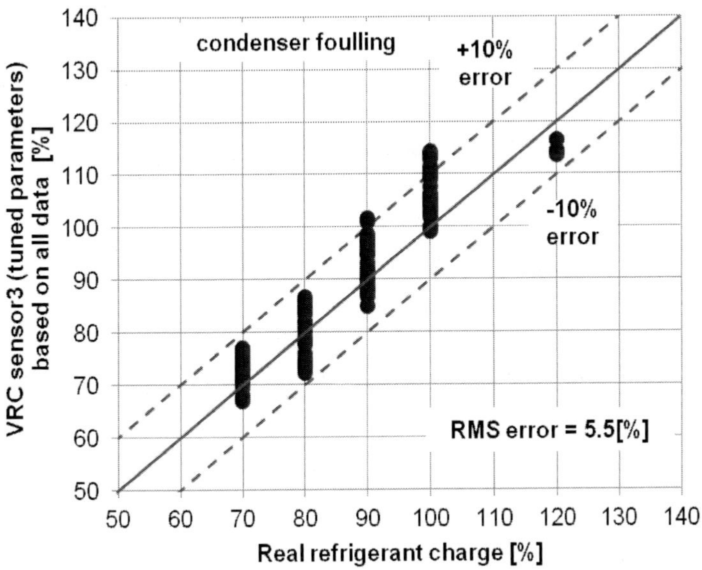

Figure 7. Performance of VRC sensor model III based on tuned parameters for RTU II under condenser fouling.

Validation of Virtual Refrigerant Mass Flow Sensor

The estimated mass flow rate and compressor power based on virtual sensors are useful indices in monitoring and diagnosing faults based on system performance with other low-cost physical sensors. Figure 9 show comparisons of VRMF sensor I predictions based on a compressor map and sensor II outputs using an energy balance under no-fault and faulty conditions. The RMS errors between VRMF models I and II were less than 10% for single faults such as condenser fouling and refrigerant charge. The errors are somewhat higher (9%) at high refrigerant charge level and low evaporator fouling, but the sensor outputs are still reasonable for this fault. The VRMF sensors also work well with multiple simultaneous fault conditions for most cases.However, for some combinations of refrigerant charge and fouling that were tested, the RMS errors between VRMF sensor I and II were over 10%.The larger errors occurred when the superheat at the compressor inlet was below 1.5°F. The incorrect compressor suction enthalpy due to a two-phase refrigerant inlet state led to the inaccurate estimations. The system was overcharged and serious evaporator fouling was applied in these situations.

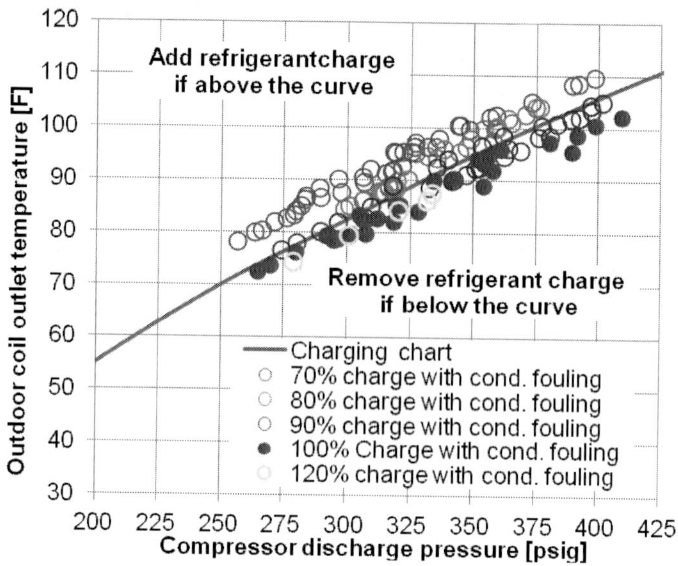

Figure 8. Charging results based on manufacturers' charging method for RTU II under heat exchanger blocking.

Figure 10 show comparisons of VRMF sensor II predictions based on the energy balance method and VRMF sensor III outputs using the TXV model both with and without faults. The RMS errors for models I and III are less than 10% and the sensors work well regardless of the fault conditions applied. Some of the larger errors may be associated with two-phase refrigerant conditions at the TXV inlet with near-zero subcooling under low refrigerant charge and condenser fouling conditions. The VRMF sensor works well under multi-fault conditions such as condenser fouling and refrigerant charge. Overall, the VRMF sensor can make accurate input power estimates regardless of faulty conditions. The output of the VRMF sensor can be used as an input of virtual performance sensor.

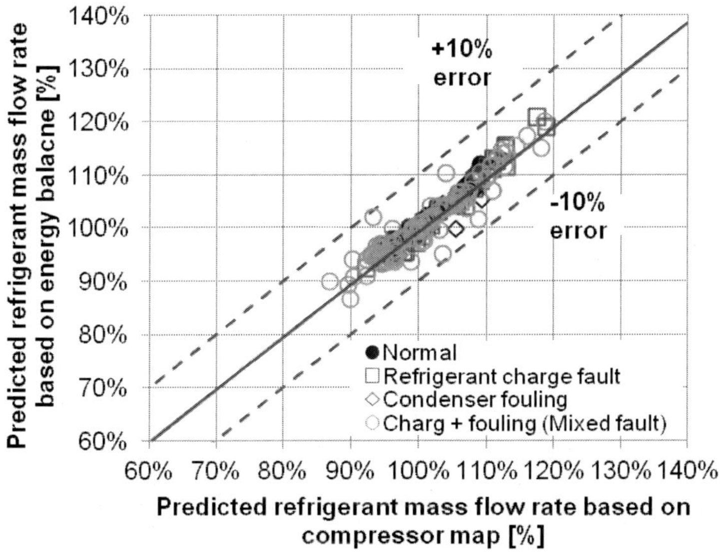

Figure 9. Performance of VRMF sensors based on model I and II under normal and faulty conditions.

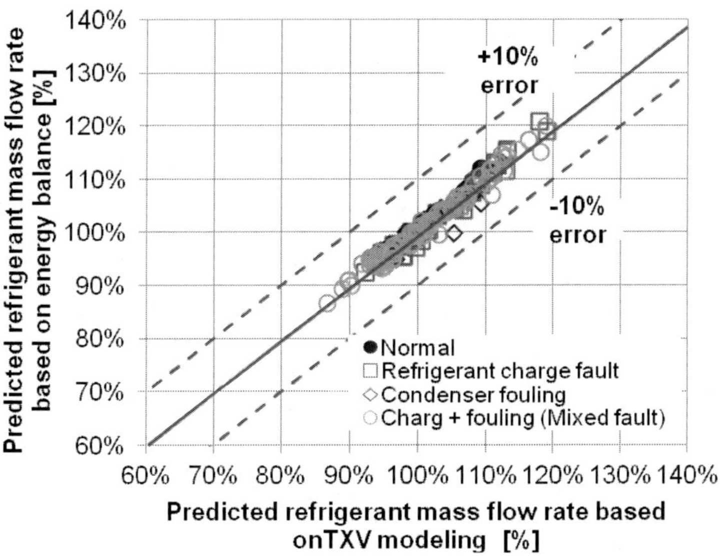

Figure 10. Performance of VRMF sensor based on model II and III under normal and faulty conditions.

Validation of Virtual Air Flow Rate Sensor

The VAF sensor for the condenser based on an energy balance was evaluated for the RTU test data.Figure 11 and Figure 12 show the condenser air flow rate estimated with the VAF sensor for the RTU systems with normal and single faulty conditions.Predicted air flow rates based on the VAF sensor can be compared to a target air flow rate to detect fouling.The target flow can be obtained from a manufacturer's catalog or from a normal value when the FDD scheme is implemented assuming that there is no fouling.

Figure 11 shows the results of the VAF sensor under 100% refrigerant charge. The VAF sensor predicted the target condenser air flow rate within 15% except under condenser fouling fault conditions. When the blocking level was over 40%, the air flow rate decreased to 65% of the nominal air flow rate.Figure 12 shows similar results but with a larger spread of predictions when the system was undercharged. As the severity of the condenser fouling increases, the estimated air flow rate is decreased.The results also show that the condenser fouling fault is decoupled from refrigerant charge faults.

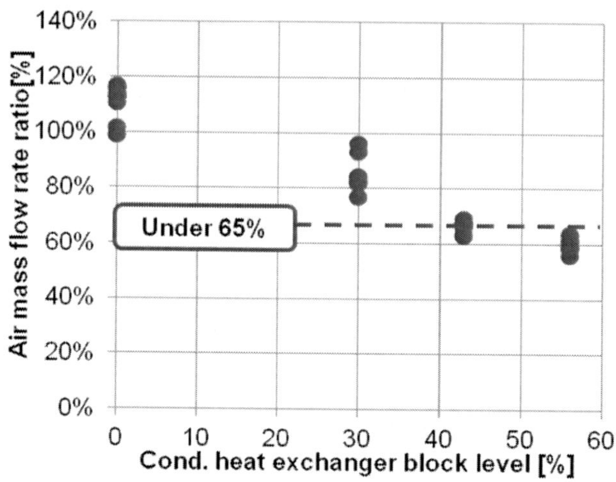

Figure 11. Performance of VAF sensor under normal refrigerant charge.

Fault Impact Model Evaluation

To evaluate the impact of faults on performance, capacity and coefficient of performance (COP) ratio were calculated. For example, the COP ratio was calculated as the ratio of estimated energy efficiency based on virtual sensors or the expected value from the fault impact model divided by the rated value provided by product specifications. The impacts of the various faults were used in deciding thresholds for the FDD demonstration system. Information about capacity, power consumption, and energy efficiency can be used in real-time monitoring of the economic status of the equipment and for decision support.

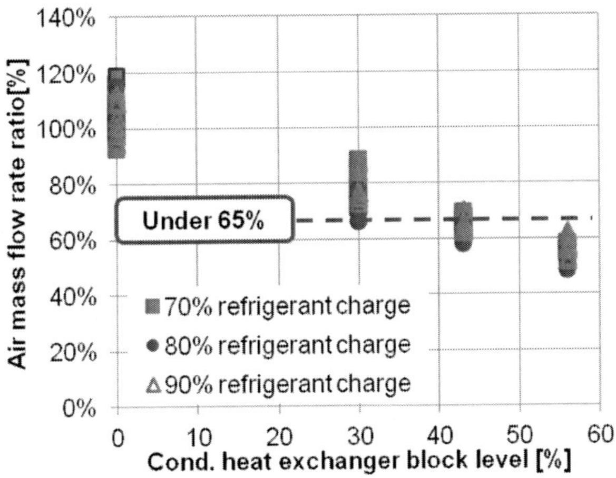

Figure 12. Performance of VAF sensor under different refrigerant charge levels.

Figure 13 shows a comparison between the virtual sensor and the fault impact model under condenser fouling fault conditions. The fault impact model matched the virtual sensor performance very well under no-fault test conditions. This means that the fault impact model can be used to estimate normal energy efficiency at faulty conditions. The energy efficiency degradation is lower than the threshold which indicates that the fault impact for condenser fouling is not sufficiently severe to require service. The COPratio decreases with increasing condenser fault levels.Although the VAF sensor indicates 40% air flow reduction, the energy efficiency was only degraded by about 7 ~ 9%. This means the potential energy savings after service is small while service costs would be relatively high.

Figure 14 shows comparisons between the virtual sensor and the fault impact model under the combination of 90% refrigerant charge and various condenser fouling conditions. When blocked heat exchanger areas were 43 and 53%, indicating condenser fouling based on the VAF sensor, it would result in over 10% energy efficiency degradation. For a 30% condenser fault level, the energy efficiency degradation was less than 10%. The service for the RTU system might be justified when the VAF sensor indicates 43% condenser fouling level. Figure 15shows comparisons between virtual sensor and fault impact model outputs under the combination of 70% refrigerant charge and various condenser fouling conditions. When the VRC sensor detects 70% as the charge level with fault detection, it can be seen that all data points are lower than the threshold for energy efficiency. This means the potential energy savings after service is small while service costs would be relatively high. When the refrigerant charge level is 70%, the current energy efficiency ratios were reduced outside the range of the thresholds. Overall, service for the RTU system might be warranted when the VRC sensor indicates 70% refrigerant charge level or less.

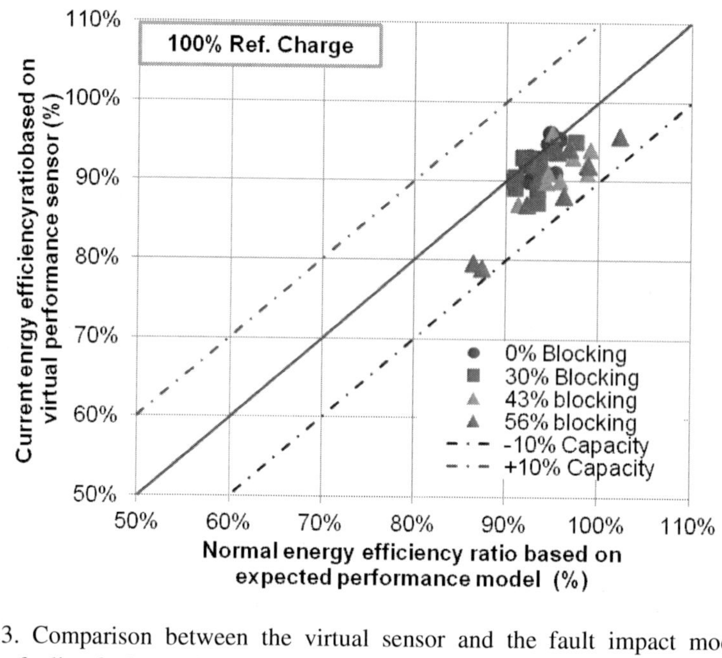

Figure 13. Comparison between the virtual sensor and the fault impact model under condenser fouling fault conditions.

Embedded FDD System Based on Virtual Sensor

This section presents the interface for a complete implementation and demonstration of the automated FDD system applied to RTU field test data. The implementation incorporates integrated virtual sensors to provide the diagnostic outputs and the performance impact of the fault(s). The demonstration system provides the following diagnostic outputs: low refrigerant charge and fouled condenser. Health and economic status reports for equipment can be generated using fault impact indices, such as system cooling capacity and COP. In particular, the fault impact indices can be used to assess the economics associated with servicing a unit if faults exist.

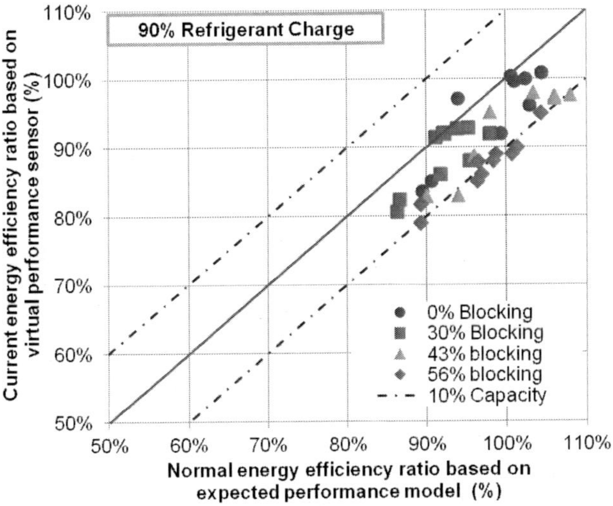

Figure 14. Comparison between thevirtual sensor and the fault impact model under 90% refrigerant charge and condenser fouling fault conditions.

User Interface Development

The Automated FDD system incorporates 1) the status of compressors and fans and the steady state detector, 2) the component fault identification information based on integrated virtual sensors, 3) health and economic status based on performance indices for on-line monitoring and 4) the complete diagnostic information about FDD status to choose proper action, shown in Figure 16.The status of fans and compressors is displayed on the left side. If steady-state conditions are detected, shown on the middle of the left side, the virtual sensors will indicate their detected values. Virtual sensors were implemented for the compressor, expansion device, condenser fouling, and refrigerant charge. The output of a virtual sensor can be compared to the output of a real sensor or other virtual sensors in order to detect and localize a fault within the system. On the right side, the system performance is also displayed to show the capacity and COP ratio. The capacity and COP ratio are the ratios of current equipment performance from virtual sensors to that of the fault impact model. The fault impact model was developed based on normal test data obtained from the manufacturer. If faults exist, the capacity and COP ratio can be used to determine whether the fault is severe enough to request service.

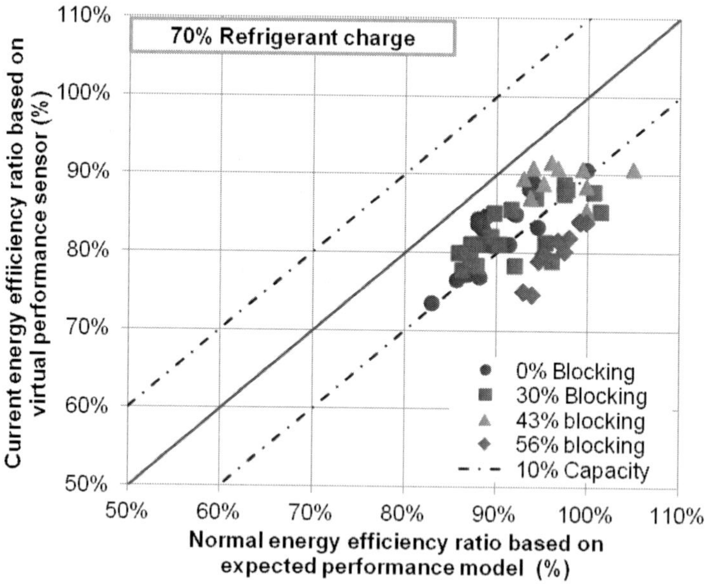

Figure 15. Comparison between the virtual sensor and fault impact model under 70% refrigerant charge and condenser fouling fault conditions.

Virtual Sensor Implementation

Figure 17 shows refrigerant charge information that is displayed by clicking the VRC sensor button. The user interface shows the current estimated level and other statistical information. It also shows the data trends as time passes. In addition, probability distributions are shown corresponding to the current estimations (red line) and expected values for normal conditions (white line). The mean and standard deviation for the current distribution is also presented in the user interface. The confidence interval is used to indicate the reliability of the current charge level from the virtual sensor. The difference between the current charge level and the expected value is used as a residual input to a Bayesian classifier in order to evaluate whether a refrigerant fault is present. The probability graph shows much larger differences between current and expected probability distributions, indicating low a refrigerant charge fault. The trending graph provides more evidence for the existence of this fault. The current and recent trending of refrigerant charge levels is shown. In addition, probability distributions are shown that correspond to the current estimations from the VRC sensor and the expected values for a normal condition. The mean and standard deviation for the current distribution is also presented. The difference between current and expected values is used as a residual input to a Bayesian classifier in order to evaluate whether a refrigerant fault is present. For the no-fault test data, there is a significant overlap between the current and expected probability distributions indicating normal operation.

Figure 18 shows the refrigerant mass flow rate information displayed on the user interface. When the compressor fault button is selected, detailed refrigerant mass flow rate information is obtained. Three VRMF sensors are implemented in the FDD system for the RTU. The virtual sensors use a compressor map and a TXV model as current predictions and an energy balance as a normal expected prediction. Outputs from the three VRMF sensors can be compared in order to detect a fault. The Bayesian classifier is applied to each combination of residuals between two predictions to identify the existence of a fault such as a compressor valve leakage or a faulty expansion device. For the no-fault test data, there is a significant overlap between the current and expected probability distributions, indicating normal operation. The capacity and COP ratios are 97% and 99% of the normal condition based on the fault impact model. Based on the readings, the diagnostic outputs indicate at the bottom right corner that the status of the RTU system is acceptable and any service is unnecessary.

Figure 19 shows the diagnostic interface based on the simulated fouling test data with 55% condenser blocking. This interface appears when the condenser fault button is selected. The condenser fouling fault can be detected by comparing the estimated air flow rate with a target value. A VAF sensor indicates the condenser air flow rate as 50% of the normal value. The probability graph shows much larger differences between current and expected probability distributions, indicating the low air flow rate fault. It is also a good feature for diagnosing fan problems.

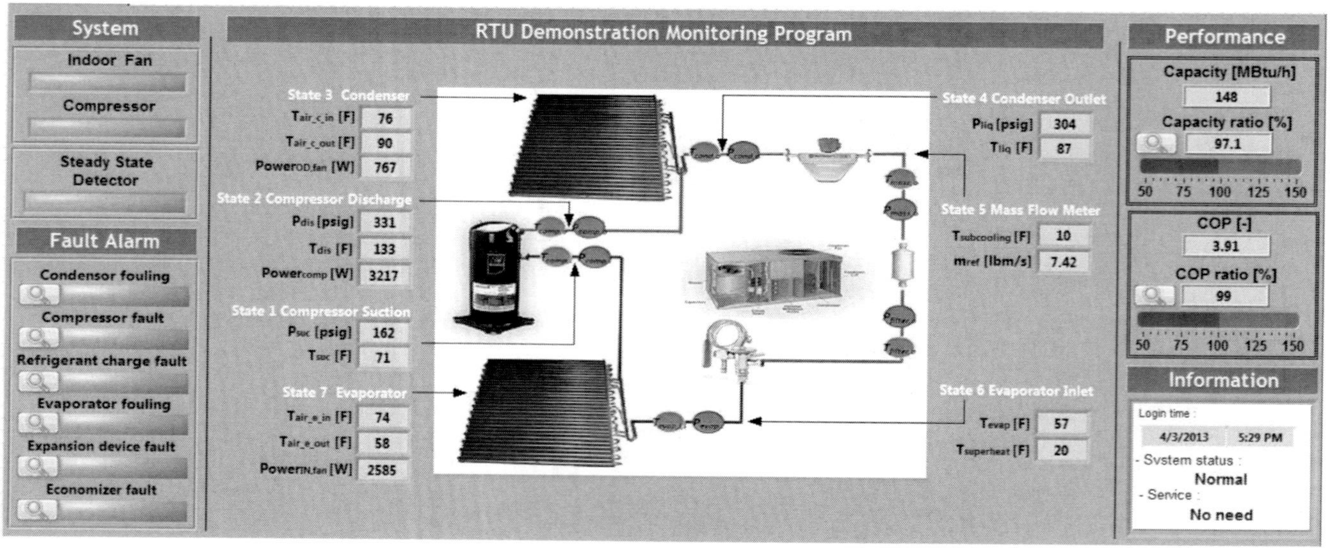

Figure 16. Implementation and demonstration of an automated FDD system for RTU under normal conditions.

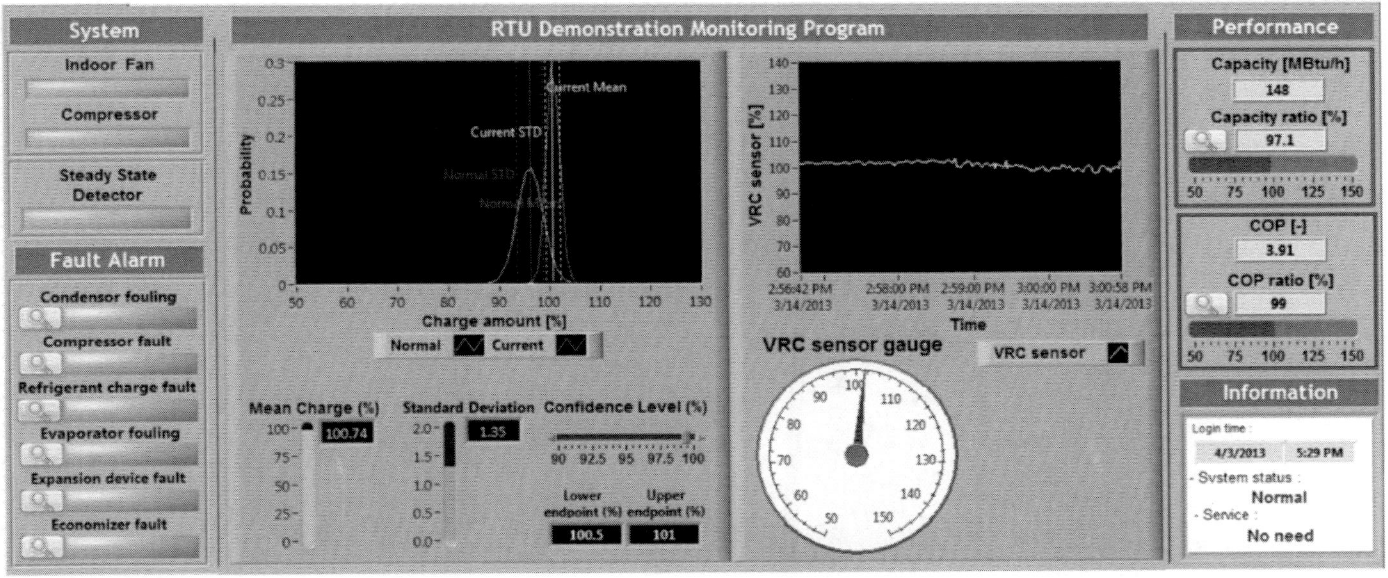

Figure 17. Implementation and demonstration of the VRC sensor under normal conditions.

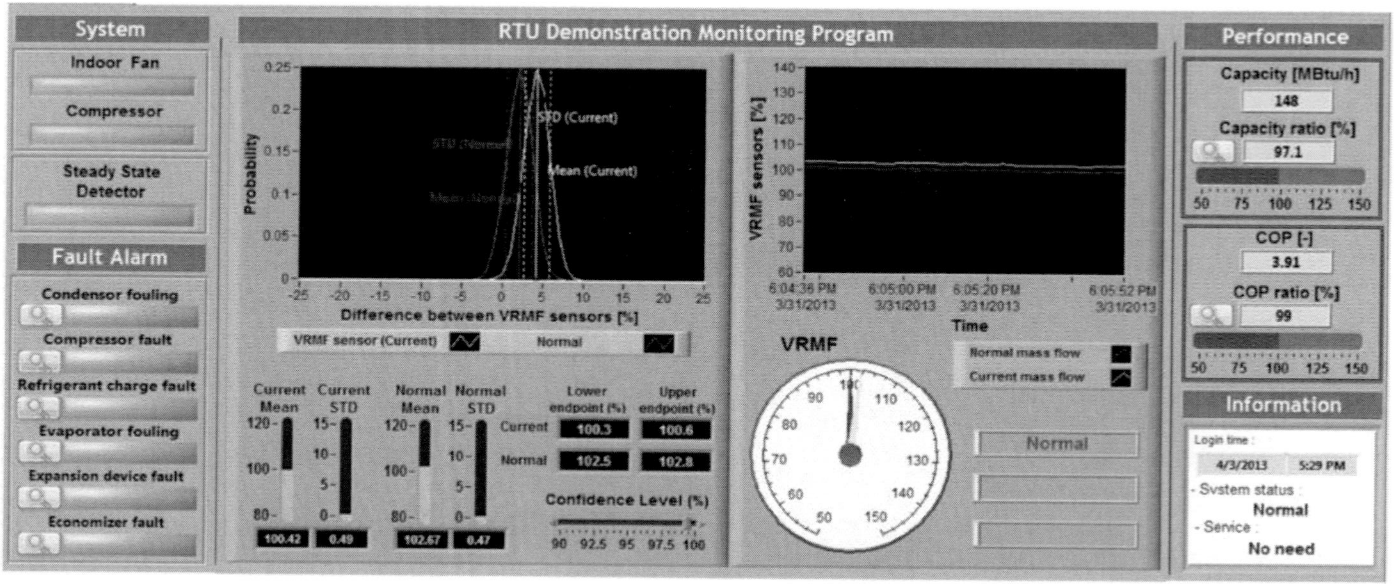

Figure 18. Implementation and demonstration of the VRMF sensor under normal conditions.

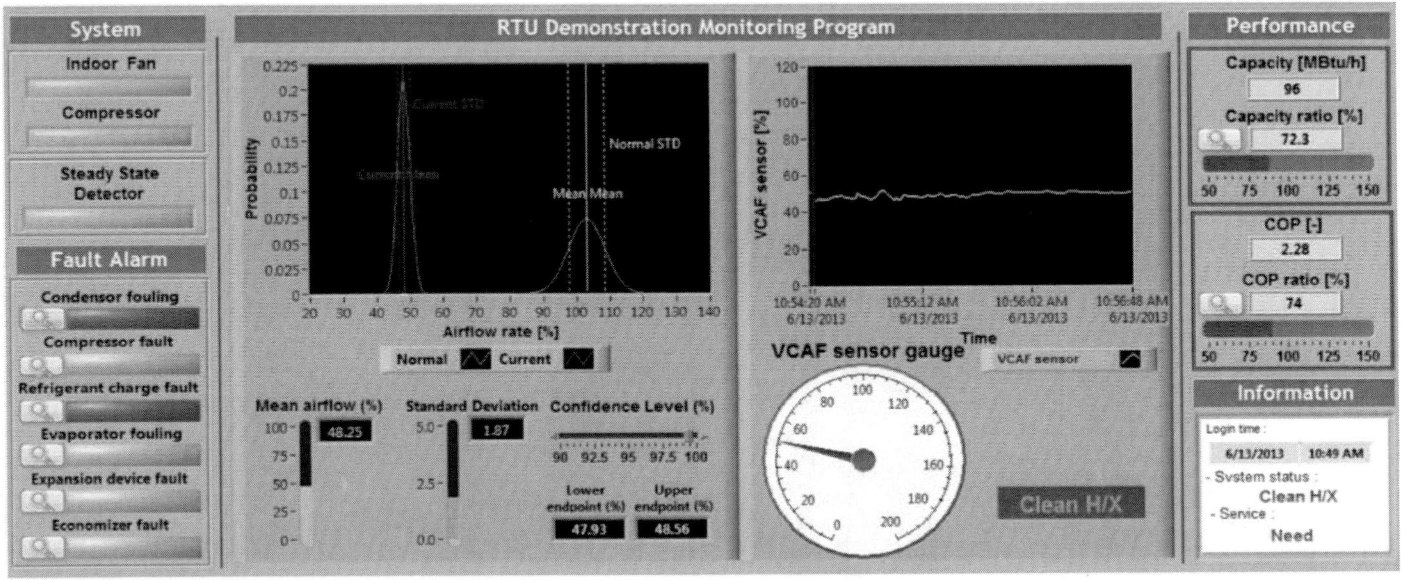

Figure 19. Implementation and demonstration of the VRMF sensor under 50% condenser blocking condition.

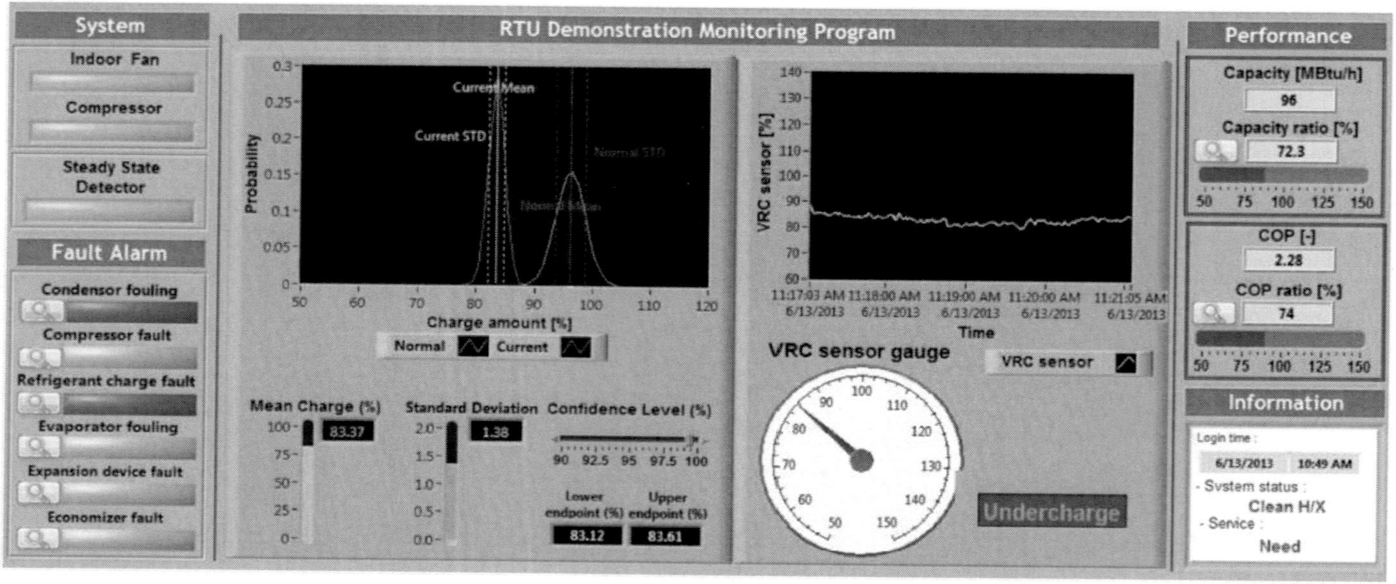

Figure 20. Implementation and demonstration of the VRC sensor under 85% refrigerant charge fault condition.

Figure 20 shows detailed refrigerant charge information based on the results with 85% refrigerant undercharge. The probability graph shows large differences between the current estimated refrigerant charge and the expected values. The trending graph provides more evidence for the existence of this fault. Overall, the demonstration of FDD system recommends that service is needed for this refrigerant undercharge fault. For this case, the capacity and COP ratio were reduced to 72% and 74%, respectively, due to low charge and condenser fouling. Based on the readings, the diagnostic outputs indicate that service is needed for this condenser fouling and refrigerant charge fault at the bottom right corner.

Conclusion

The primary goal of this research is to develop and demonstrate an integrated, on-line performance monitoring and diagnostic system with low cost sensors for RTU. Automated FDD has the potential for improving energy efficiency along with reducing service costs and comfort complaints. The FDD system was implemented and demonstrated for the RTU that incorporates integrated virtual sensors and fault impact evaluation for decision support. To achieve this goal, virtual sensors with low cost measurements and simple models were developed to estimate quantities that would be expensive and or difficult to measure directly. virtual sensors use manufacturers or existing test data to determine the parameter.So, it is easy to be customized for each system. Various virtual sensors were implemented for the compressor, TXV as expansion valve, condenser heat exchangers, and refrigerant charge in order to provide the following diagnostic outputs. The tests also quantified the benefits of this technology with measurements of equipment performance and demonstrated implementation with low sensor costs. Once faults are detected and the causes of the faults are identified, an assessment of the severity of a fault is essential to the decision process and virtual sensors can be used as inputs to this analysis. Health and economic status reports for equipment can be generated using fault impact indices, such as capacity and energy efficiency performance reduction ratios. In particular, the fault impact indices can be used to assess the economics associated with servicing a unit if faults exist.

References

ADM, 2009, Market "Assessment and field M&V Study for Comprehensive Packaged A/C Systems Program," Final Report.

ANSI/ARI Standard 540-99, 2004, "Standard for Performance Rating of Positive Displacement Refrigerant Compressors and Compressor Units," Air-Conditioning, Heating, and Refrigeration Institute, Arlington, VA 22201, US.

Braun, J.E., 2003, "Automated Fault Detection and Diagnostics for Vapor Compression Equipment," *ASME Journal of Solar Energy Engineering*, Vol. 125, Page 266-274.

Cowan, A., 2004, "Review of recent commercial rooftop unit field studies in the Pacific Northwest and California." Northwest Power and Conservation Council and Regional Technical Forum, Portland, Oregon.

Daily, M., J. Harris, D. Keirsey, D. Olin, D. Payton, K. Reiser, J. Rosenblatt, D. Tseng, and V. Wong. "Autonomous cross-country navigation with the ALV." In: *Robotics and Automation*, 1988. Proceedings, 1988 IEEE International Conference on, pp. 718-726. IEEE, 1988.

Dixon, Kevin, John Dolan, Wesley Huang, Christiaan Paredis, and Pradeep Khosla. "RAVE: A real and virtual environment for multiple mobile robot systems." In: *Intelligent Robots and Systems*, 1999. IROS'99. Proceedings. 1999 IEEE/RSJ International Conference on, vol. 3, pp. 1360-1367. IEEE, 1999.

Gustafsson, F, 1998, "Estimation and Change Detection of Tire-Road Friction Using the Wheel Slip," *IEEE Control System Magazine*, Vol.18, No.4, Page 42-49.

Heredia, G and Ollero, A., 2010, "Virtual Sensor for Failure Detection, Identification and Recovery in the transition Phase of a Morphing Aircraft," *Sensors*, Vol.10, No.3, Page 2188-2201.

Healy, W., 2010, "Vapor Compression Equipment Sensors and Energy Monitoring Systems, Vapor Compression Equipment America Meeting on Diagnostic Measurement and Performance Feedback for Residential Space Conditioning Equipment," National Institute of Standards and Technology, Gaithersburg, MD.

J. Stéphant, A. Charara, and D. Meizel, 2004, "Virtual Sensor: Application to Vehicle Sideslip Angle and Transversal Forces," *IEEE Transactions on Industrial Electronics*, vol. 51, no. 2, pp. 278-289, April 2004.

Kabadayi. S, Pridgen. A, and Julien. C, 2006, "Virtual Sensors: Abstracting Data from Physical Sensors, International Symposium on a World of Wireless," *Mobile and Multimedia Networks*, Page 587-592.

Katipamula, S. and Brambley, M. R., 2005a, "Methods for Fault Detection, Diagnostics, and Prognostics for Vapor Compression Equipment Systems, A Review, Part I," *HVAR&R Research*, Vol. 11, No. 1, Page 3-25.

Katipamula, S. and Brambley, M. R., 2005b, "Methods for Fault Detection, Diagnostics, and Prognostics for Vapor Compression Equipment Systems, A Review, Part I," *HVAR&R Research*, Vol. 11, No. 1, Page 3-25.

Kim, W., 2013. Fault detection and diagnosis for air conditioners and heat pumps based on virtual sensors. Ph.D. Thesis, Purdue University, IN.

Kim, W. and Braun, J. E., 2012a, "Performance Evaluation of a Virtual Refrigerant Charge Sensor," *International Journal of Refrigeration.*

Kim, W. and Braun, J. E., 2012b, "Evaluation of the Impact of Refrigerant Charge on Air conditioner and Heat pump performance," *International Journal of Refrigeration*, Vol.35, No. 7, Page 1805-1814.

Kim, W. and Braun, J. E., 2016, "Development and evaluation of virtual refrigerant mass flow sensors for fault detection and diagnostics."*International Journal of Refrigeration* Vol. 63, Page184-198.

Kumar, R., Shin, J., Iftode, L. and Rama Chandran, U., 2008, "Mobile Virtual Sensors: A Scalable Programming and Execution Framework for Smart Surveillance," In: *Proceedings of the 5th Workshop on Embedded Network Sensors.*

Li, H. and Braun, J. E., 2007a, "Decoupling Features and Virtual Sensors for Diagnosis of Faults in Vapor Compression Air Conditioners." *International Journal of Refrigeration*, Vol. 30, No. 3, Page 546-564.

Li, H. and Braun, J. E., 2007a, "Decoupling Features and Virtual Sensors for Diagnosis of Faults in Vapor Compression Air Conditioners." *International Journal of Refrigeration*, Vol. 30, No. 3, Page 546-564.

Li, H. and Braun, J. E., 2007b, "A Methodology for Diagnosing Multiple-Simultaneous Faults in Vapor Compression Air Conditioners."*HVAC&R Research*, Vol. 13, No.2, Page 369-395.

Li, H. and Braun, J. E., 2009a, "Development, Evaluation, and Demonstration of a Virtual Refrigerant Charge Sensor." *HVAC&R Research*, Vol. 15(1), Page 117-136.

Li, H. and Braun, J. E., 2009b, "Virtual Refrigerant Pressure Sensors for Use in Monitoring and Fault Diagnosis of Vapor-Compression Equipment," *HVAC&R Research*, Vol. 15, No. 3, Page 597-616.

Li, H. and Braun, J. E., 2009c, "Virtual Refrigerant Pressure Sensors for Use in Monitoring and Fault Diagnosis of Vapor-Compression Equipment," *HVAC&R Research*, Vol. 15, No. 3, Page 597-616.

Lichuan Liu., Kuo, S. M., and Meng Chu Zhou., 2009, "Virtual Sensing Techniques and their Applications," *International Conference on Networking, Sensing and Control*, Page 31-36.

New Buildings Institute, 2003, "Integrated Energy Systems: Productivity and Building Science," Available online: http://www.energy.ca.gov/pier/project_reports/.

Messenger, M., 2008, "Strategic Plan to Reduce the Energy Impact of Air Conditioners," CEC-400-2008-010.

Rallo, R., Ferre-Gine, J., and Arenas, A., 2002, "Neural virtual sensor for the inferential prediction of product quality from process variables," *Computers and Chemical Engineering*, Vol. 26, Page 1735-1754.

Wichman, A. and Braun. J. E., 2009, "A Smart Mixed-Air Temperature Sensor," *HVAR&R Research*, Vol. 15, No.1, Page 101-115.

Wilson, Edward. "Virtual sensor technology for process optimization." In: *Symposium on Computers and Controls in the Metals Industry in Iron and Steel Society*. 1997.

Yu, D., Li, H. and Braun, J. E., 2011, "A Review of Virtual Sensing Technology and Application in Vapor compression equipment Systems." *HVAC&R Research*, Vol. 17, Page 619-645.

Biographical Sketch

Woohyun Kim

Affiliation: Pacific Northwest National Laboratory

Education:

- PHD Mechanical Engineering, Purdue University
- MS Mechanical Engineering, Purdue University
- BS Sung Kyun Kwan University, Seoul, South Korea

Research and Professional Experience

9/13 – Present: Engineer Efficiency Engineer, Electricity Infrastructure and Buildings, Pacific Northwest National Laboratory, Richland, WA, US

5/02 - 05/07: Senior Researcher, Digital Appliance R&D Center, Samsung Electronics, Co., Ltd. Suwon, Korea

3/00 - 4/02: Researcher, R&D Center, Mando Climate Corporation, Seoul, Korea

Professional Appointments

- Reviewer for International journal of refrigeration
- Reviewer for International Refrigeration and Air Conditioning Conference

Honors

- Top paper award of 2012/2013 academic year's Very Highly Commended Paper by the International Journal of Refrigerant (IJR) (2013).
- First place Science Award at by Energy Efficient Buildings Hub (EEB Hub) spring conference graduate student poster sponsored by the US Department of Energy (2013).

Publications

- Kim W, and J Braun. 2016. "Development and evaluation of virtual refrigerant mass flow sensors for fault detection and diagnostics." *International Journal of Refrigeration.*
- Katipamula S, W Kim, RG Lutes, and RM Underhill. 2015. Rooftop Unit Embedded Diagnostics: Automated Fault Detection and Diagnostics (AFDD) Development, Field Testing and Validation. PNNL-23790, Pacific Northwest National Laboratory, Richland, WA.
- Kim W, and J Braun. 2015. "Extension of a Virtual Refrigerant Charge Sensor." *International Journal of Refrigeration.*
- Kim, W. H. and Braun, J. E. (2013), Performance Evaluation of a Virtual Refrigerant Charge Sensor, *International Journal of Refrigeration.*

INDEX

I

J

L

M

N

O

P